T. W. Schultz

Successful Seed Programs

IADS DEVELOPMENT-ORIENTED
LITERATURE SERIES
Steven A. Breth, series editor

Successful Seed Programs
was prepared under the auspices of
the International Agricultural Development Service
with partial funding from
the German Agency for Technical Cooperation
(Deutsche Gesellschaft für Technische Zusammenarbeit)

ALSO IN THIS SERIES

*Rice in the Tropics: A Guide to the Development
of National Programs*, Robert F. Chandler, Jr.

*Small Farm Development: Understanding and Improving
Farming Systems in the Humid Tropics*, Richard R. Harwood

Successful Seed Programs
A Planning and Management Guide

compiled and edited by
Johnson E. Douglas

Westview Press / Boulder, Colorado

Copyright © 1980 by the International Agricultural Development Service

Published in 1980 in the United States of America by
  Westview Press, Inc.
  5500 Central Avenue
  Boulder, Colorado 80301
  Frederick A. Praeger, Publisher

Library of Congress Cataloging in Publication Data
Main entry under title:
Successful seed programs.
    (IADS development-oriented literature series)
    Bibliography: p.
    1. Seed industry and trade. 2. Seed distribution. I. Douglas, Johnson E. II. Series: United States. International Agricultural Development Service. IADS development-oriented literature series.
SB114.A3S92   631.5'21   79-28308
ISBN 0-89158-793-4
ISBN 0-89158-872-8 pbk.

Printed and bound in the United States of America

Dedicated to my wife, Luciene,
who has contributed in countless ways
to making this book a reality rather than a dream

Contents

Figures

Tables

Foreword

A seed program capable of providing farmers with good quality seed is essential to a nation's agricultural development. The farmer is the user of the product, and by his perceptions he is the one who gives it the final and most rigorous test. The farmer, as seed grower, is also the source of the product. Perhaps no other industry is as closely involved with the development process at the farm level as is the seed industry.

The most appropriate form for a seed program may differ widely from one country to another. To be effective the seed program must be carefully tuned to the nature of farming in the country—that is, to the level of sophistication of the agriculture and its stage of development. Thus, a seed program might supply quite different services in isolated mountain valleys within a country than it would provide in the plains or other more accessible places. Over time, the seed program must also be prepared to change. What is appropriate in the early stages of development may not be appropriate a few years later. On the other hand, the introduction of a too sophisticated program may lead to serious disappointments because of the inability of other aspects of the agricultural economy to keep up.

The seed program must also interact closely with the nation's research system. It must be able to move new varieties to the farmer promptly and efficiently. All too often the plant breeders have a new variety ready for release before earlier ones have made their way through the complex stages of multiplication and distribution that lie between the experimental field and the farms of a nation. Cooperation is, of course, a two-way street, and it is incumbent upon the researcher to work with the seed industry.

This book incorporates the work of an exceptionally large number of contributors and the comments of a long list of reviewers. The result is perhaps a broader coverage of the principles of successful seed programs than would have been possible by any other route. It is not a cookbook offering easy outlines for establishing a seed program. Rather, it is a source of ideas and principles that may be put together to build and improve a seed program.

Johnson E. Douglas did the arduous work of coordinating, compiling, organizing, reconciling, and rewriting. Douglas has many years of experience in seed programs in Tunisia, India, and the United States of America as well as being a frequent consultant to programs in several other countries. He is currently seed specialist at the Centro Internacional de Agricultura Tropical (CIAT) in Colombia.

This book is part of the IADS series of development-oriented literature. We are grateful to the German Agency for Technical Cooperation (Deutsche Gesellschaft für Technische Zusammenarbeit) and the Rockefeller Foundation for providing funds to develop the book.

A. Colin McClung
Acting President
International Agricultural Development Service

Preface

A seed program must be understood for what it is—a "service" to farmers. The world's farmers, not governments nor private companies, produce the food for all consumers. Seed programs fulfill their service function only when millions of farmers are able to obtain and plant better seed.

In the past, emphasis has often been placed on certain segments of seed programs. Now, more attention is being given to the total seed program needs of developing countries. Though external assistance has helped and can help in the future, the building of a total seed industry in a country is basically a "do-it-yourself" activity. Much technical information is available about how to grow, process, test, and store seed. However, very little is written about questions that face the administrators, managers, and other leaders of seed programs on how to plan, implement, and manage seed program activities.

The International Agricultural Development Service has prepared this book to meet that need. Experiences from both developed and developing areas have been utilized to guide administrators, managers, and leaders. An attempt has been made to concentrate on *essential* issues and aspects of programs that will be most useful in developing nations today.

The specific purposes of this book are (1) to provide the administrator with a basic reference for assessing the stage of development of the seed program, identifying needs, establishing priorities, developing plans, and taking actions necessary to build a successful seed program and (2) to help managers and leaders of the different aspects of a seed program to plan and implement successful activities that will result in enhanced seed supplies. The book does not attempt to provide all the technical information needed to carry out the individual activities; publications that can be utilized for this purpose have been included in the Bibliography.

Nine groups of people in developing countries are considered the main audience for this book: (1) planners and administrators in government concerned with seed program development and seed supplies, (2) leaders of national research programs and projects, (3) directors of extension and information services and information specialists primarily concerned with seed, (4) leaders of initial seed increase projects or enterprises, (5) managers of seed enterprises and heads of production and marketing programs, (6) heads and managers of seed quality control activities, (7) seed technologists interested in assuming more responsibility in seed programs, (8) instructors in seed technology educational and training programs, and (9) donor agencies and expatriate seed technologists.

The book was prepared by sixteen internationally known consultants who worked in pairs on individual chapters with a coordinating editor. The experiences of the contributing consultants encompass seed programs in developed and developing countries. Subsequently, fifty experienced leaders, technologists, and administrators in seed programs, as well as the contributing consultants, read and commented on an early draft. As a result, this volume reflects the broad views of the international seed community. Nevertheless, in the process of reconciling certain divergent viewpoints and integrating different writing styles and emphases, some errors or omissions may have occurred. If so, the responsibility is mine.

The authors of the book recognize that countries are in different stages of development, some are larger than others, and their political backgrounds differ. No one country can use all proposals and alternatives. Choices need to be made.

Chapters 2, 3, 4, 5, and 6 cover individual aspects of a seed program. The other chapters deal with topics that are common to all components. It is, therefore, necessary to consider the book as a total unit. Many appendixes, containing valuable information that is too lengthy or too detailed to include in the body of the text, have been included. The Glossary is provided because several terms have very specific meanings. The Bibliography lists a wide range of references that may be useful.

Those contributing to this book hope that it will help decision makers reach more meaningful decisions today and build more successful seed programs tomorrow.

Johnson E. Douglas
Cali, Colombia

Acknowledgments

Many persons and organizations contributed to this work; without their help the book would not have been possible. Special acknowledgments go to the contributing consultants, who gave of their time, experience, and ability, and to their organizations, which supported them while they worked on the project.

H. DEAN BUNCH

Director, Office of International Programs, Mississippi State University, United States of America

FRANCIS C. BYRNES

Program Officer, International Agricultural Development Service, United States of America

DAVID L. CURTIS

Director, International Seed Operations, DeKalb Ag Research, Inc., United States of America

JAMES C. DELOUCHE

In Charge, Seed Technology Laboratory, Mississippi State University, United States of America

GEORGE M. DOUGHERTY

Seed Processing Specialist, Mississippi State University, United States of America

WALTHER P. FEISTRITZER

Coordinator, Seed Improvement and Development Program, FAO, Italy

WAYNE H. FREEMAN

Leader, Integrated Cereals Project, International Agricultural Development Service, Nepal

ALEXANDER GROBMAN

Director, International Cooperation, Centro Internacional de Agricultura Tropical, Colombia

LENNART KAHRE
Head, Swedish Seed Testing and Certification Institute, Sweden

A. FENWICK KELLY
Deputy Director, National Institute of Agricultural Botany, United Kingdom

R. L. PALIWAL
Associate Director, International Maize Program, Centro Internacional de Mejoramiento de Maíz y Trigo, Mexico

J. M. POEHLMAN
Professor of Agronomy, University of Missouri, United States of America

HOWARD C. POTTS
Professor of Agronomy, Mississippi State University, United States of America

S. F. ROLLIN
Commissioner, Plant Variety Protection Office, U.S. Department of Agriculture, United States of America

OLLE SVENSSON
Seed Specialist, Swedish Seed Testing and Certification Institute, Sweden

CARLOS VECHI
IPB Comercio de Sementes Ltda, Brazil

The time and numerous suggestions given by the many reviewers is also sincerely appreciated. Persons who commented on the circulated draft were:

S. Abu-Shakra, University of California, United States of America
Primo Accatino, CIP, Turkey
Bjorn Almquist, Jorgen Ankergaton, Sweden
R. Glenn Anderson, CIMMYT, Mexico
C. Anselme, Groupe d'étude et de contrôle des variétés des semences, France
Charles Baskin, Mississippi State University, United States of America
Russell H. Bradley, Experience Incorporated, South Korea
W. T. Bradnock, Production and Marketing Branch, Canada
James Bryan, CIP, Peru
A. S. Carter, United States of America
J. Ritchie Cowan, IRRI, Philippines

Wendell P. Ditmer, Pennsylvania Department of Agriculture, United States of America

L. E. Everson, Iowa State University, United States of America

John E. Ferguson, CIAT, Colombia

Eleodoro J. Fuentes, Seminarios Panamericanos de Semillas, Chile

Richard C. Gartrell, Harvard University, United States of America

William G. Golden, Jr., Hawaiian Agronomics Co. (International), Solomon Islands

Donald Grabe, Oregon State University, United States of America

James F. Harrington, University of California, United States of America

M. N. Harrison, United Kingdom

R. L. Harty, Department of Primary Industries (Queensland), Australia

J. Robert Huey, United States of America

Robert Jacobsen, Danske Landboforeningers Frøforsyning, Denmark

Louisa A. Jensen, (Retired) Oregon State University, United States of America

M. S. Joshi, FAO, Sudan

David Juckes, OECD, France

Robert W. Jugenheimer, University of Illinois, United States of America

Benicjusz Kramski, Ministerstwo Rolnictwa, Poland

Milton W. Lau, USAID, Peru

H. H. Leenders, Fédération Internationale du Commerce des Semences, Netherlands

Earl Leng, USAID, United States of America

Harold D. Loden, American Seed Trade Association, United States of America

Robert A. Luse, CIAT, Colombia

W. L. McCuistion, Oregon State University, United States of America

D. B. Mackay, National Institute of Agricultural Botany, United Kingdom

J. D. Maguire, Washington State University, United States of America

P. Maleki, FAO/World Bank, Italy

Heribert Mast, Union Internationale pour la Protection des Obtensions Végétables, Switzerland

P. R. Mezynski, Texas State Technical Institute, United States of America

Delbert T. Myren, USAID, United States of America

P. H. Nelson, Kleinwanzlebener Saatzucht Ag, West Germany

Robert D. Osler, CIMMYT, Mexico

Joseph K. Park, U.S. Department of Agriculture, United States of America

D. C. Pickering, World Bank, United States of America

Erlinda Pili-Sevilla, Department of Agriculture, Philippines

E. H. Roberts, University of Reading, United Kingdom

S. Sadjad, Bogor Agricultural University, Indonesia

W. O. Scott, University of Illinois, United States of America

S. M. Sehgal, Pioneer Hi-Bred International, Inc., United States of America

L. E. Smith, Society of Commercial Seed Technologists, United States of America

M. S. Swaminathan, Indian Council of Agricultural Research, India

J. R. Thomson, Edinburgh School of Agriculture, United Kingdom

Jack D. Traywick, IADS, Panama

W. H. Verburgt, Kenya Seed Company, Kenya

Eduardo Zink, Instituto Agronômico de Campinas, Brazil

The coordinating editor is especially indebted to Luciene Douglas, his wife, who not only typed many drafts but also contributed to the editing of the manuscript. Her encouragement and assistance benefited the project much more than a reading of the book reveals.

Several people at the Centro Internacional de Agricultura Tropical, Cali, Colombia, have contributed much through their secretarial, technical, and artistic talents. The final artwork was done by Fanny Rodriguez, Oscar Idarraga, and Gerardo Gonzalez.

Staff members and secretaries of IADS have provided support and assistance in too many ways to mention, but the total support provided could not have been better. Steve Breth, editor for this series of IADS publications, has contributed immensely to the book, especially by improving its readability and clarity. The Rockefeller Foundation's willingness to allow me to contribute time to this project is also acknowledged and immeasurably appreciated.

If ever a book was prepared by a true team effort, this one achieved that goal. It is hoped that the spirit of cooperation that permeated this project will spread into the development of effective seed programs in country after country.

J.E.D.

Successful Seed Programs

Overview:
Successful Seed Programs in Brief

The test of an administrator or a leader is the ability to reach the correct decision at the right time. Leaders of seed programs must make many hard choices and establish consistent policies so that seed production and distribution accelerate agricultural progress rather than impede it. Good decisions can be immensely beneficial to the farmer and the nation; incorrect decisions can be disastrous.

What are the elements of a successful seed program? How can they be assembled and integrated? What are the priorities? Ways of assessing many options and choosing among them are explained in detail in the following chapters. *This introduction is intended to summarize the principal administrative points, chapter by chapter, and to provide an overview of the organizational interrelationships of a nation's seed activities.*

MAJOR ADMINISTRATIVE POINTS

Seed Supplies Today and Tomorrow (Chapter 1)

Seed is not *just* something planted by farmers. It is the carrier of the genetic potential for higher crop production. Seed of improved varieties can trigger change and help bring about agricultural production objectives. For seed to be a catalyst, however, the seed supply pipeline must flow freely. Policy decisions at high levels are required to ensure that seed supplies and seed program development receive the priorities they deserve.

Often, to establish the basis for sound decisions, the current situation should be reviewed and assessed. To do so, many countries form a seed review team. A major objective of such a team should be to determine the stage of development of the nation's seed program or activities. Typical stages of seed program development are

Stage 1. A plant breeding department is multiplying small quantities of seed and distributing it to a very few farmers.

1

Stage 2. Seed is being multiplied by the plant breeding department, but distribution is to selected seed growers who multiply it further. Nevertheless, little improved seed is on the market.

Stage 3. The nation has a policy for development of a seed program; and seed production, marketing, quality control, certification, and training are in operation.

Stage 4. National seed policy has been reexamined; attention is being given to developing and strengthening commercial seed production and marketing; a seed law is in effect; training is continued; and links with many related institutions and groups are being established.

A seed program in stage 1 may be adequate for meeting a country's short-term objectives but not for long-term needs. Improvement can start at whatever stage the program is in. Often, as programs reach stages 3 or 4, some program components are thriving while others are feeble. Each component must be evaluated to identify strengths and weaknesses. In a successful program all components of the total seed industry exist and are functioning well.

The team should review: (1) the broad agricultural development strategy in relation to seed supply needs; (2) the capabilities, objectives, and results of crop improvement research and development; (3) the policies and procedures used in initial seed increases and the quantity and quality of seed produced; (4) the capacity to build commercial seed and Certified Seed supplies; (5) the effectiveness of quality control measures; (6) the programs and activities involved in getting seed to the farmer; (7) the physical, human, financial, and external resources available; (8) the supply and distribution of inputs other than seeds and credit availability; (9) the effect of government policy on the growth of the seed industry; and (10) the various components of the seed program to see how well they are working together.

The information gathered needs to be assessed and interpreted in relation to (1) the stage of development of the nation's agriculture, (2) the present scope of seed production and supply operations, (3) the quantities of seed needed, (4) the need for balanced development of the total seed program, (5) the facilities and equipment required by the public and private sectors, (6) the adequacy of funds for capital investments or working capital, (7) the needs for seed legislation and quality control, (8) the organizational structure needed, and (9) the development of management capabilities and personnel.

As a result of the assessment, objectives for each component of the seed program should be well defined, and a phased plan for achieving the goals should logically follow. Establishing a coordinating mechanism such as a national seed board is often a good way to link the many components of a seed program and to provide the means, the will, and the essential policy decisions for sustained development.

Many policies need to be established during stages 2 and 3. The development of policies can clarify a government's position and be the basis for the systematic implementation of a properly formulated plan. Political leaders and administrators, working with a national seed board, can make a major contribution to a seed program by placing a high priority on forming well-conceived policies. Policies should be neither inflexible nor erratic. (Points on which policy decisions may be needed are listed in each chapter.)

The Genesis: Crop Breeding Research (Chapter 2)

Crop research is the foundation of a seed program. How varieties were developed, where they originated, and the source of seed from which they came are secondary matters (although not to be ignored because they are important). In terms of the farmer and a nation's agricultural production, the primary focus must be on the availability of the best varieties possible. The farmers' motivations and reasons for accepting varieties and the ease of seed multiplication must also be considered when deciding which varieties to propose for use. If the varieties perform well and are accepted by farmers, this part of the program can be considered successful.

Among the most important issues are (1) the emphasis given to breeding and testing new varieties relative to testing only varieties from outside the country for possible introduction, (2) the crop research program's success in developing varieties that have an impact on production, (3) the extent of cooperation with regional, international, and private plant breeders, and (4) the kind of testing done, who is responsible for it, and the steps taken to decide which varieties are used.

In deciding whether to breed new varieties and test them or to concentrate instead on testing varieties introduced from abroad, the economic and political importance of the crop must be considered as well as whether special factors limit the use of varieties from outside sources. For example, unusual local growing conditions may make foreign varieties unsuitable, or traditional export markets may demand varieties that have special quality characteristics. Either situation would justify a plant breeding program within a country. The adequacy of human, financial, and physical resources to start and sustain a program in public or private institutions also needs evaluation. The allocation of resources often determines the rate at which results are achieved. The role of domestic and foreign commercial seed enterprises in meeting the demand for seed of some crops should be considered. Ultimately, a careful analysis of the alternatives should reveal a way to obtain optimum results with the resources available. If breeding of one or more crops is under way, short-, medium-, and long-term goals must be clearly identified.

Crop research must be a team effort within a program, among related programs, and among institutions. It requires a competent staff, executing a program well, over a sustained period of time. Regular evaluation of a breeding program's results can help the program achieve its goals, the most important of which is having its varieties reach farmers.

In both crop breeding research and the testing of varieties introduced from outside the country, collaboration among countries that have ecologically similar conditions can be advantageous. The free exchange of germplasm and materials for testing are common types of collaboration. Also, utilizing the information and germplasm available from international agricultural research centers can strengthen breeding as well as testing programs.

Private crop breeding makes valuable contributions to the seed programs of many countries. The degree of initiative assumed by a private seed enterprise depends largely upon the commercial atmosphere within a country. The absence of excessive restrictions on imports of breeding stock and seed for multiplication will encourage the development of private seed enterprises.

Testing is conducted as part of crop breeding research to compare new varieties or as a way to evaluate varieties from outside the country. The initial testing of experimental varieties is the duty of the crop research program regardless of whether it is public or private. Most crop research programs conduct on-farm testing because of the benefits to be gained from having evaluations in several farmers' fields. The stage of development of the program affects the nature of performance testing. Tests for identity, uniformity, and stability can be useful, but in newer seed programs they are of less value, especially when most of the research is in the public sector.

Test results must be organized and made available so that extension personnel, people with seed enterprises, and marketing groups as well as farmers can assess varieties wisely. The ways of using the test results range from just providing information on the performance of varieties to developing lists of varieties that must be certified.

To harmonize the interests of the various crop research programs and to provide a mechanism for the orderly introduction of new varieties, especially from public research activities, a variety review and release committee should be created. Private seed enterprises involved in plant breeding need their own variety release mechanisms. Various approaches can be used to relate the test results from private enterprises to official tests or assessments. As regional and international cooperation in plant breeding expands, the simultaneous or joint release of varieties by two or more countries offers advantages.

The Seed Program Starts:
Initial Seed Multiplications (Chapter 3)

Although crop research is the foundation of a good seed program, it is not the program. The initial seed multiplications start seed on the way from the research station to the farmer. Multiplication and the subsequent maintenance of each variety must be done painstakingly to safeguard the genetic identify and purity of the variety. The seed of a new variety should be multiplied quickly so it can be used soon after it has been developed. Maintenance and multiplication of varieties are other ways countries working together in crop research can cooperate. Practical help often can be obtained from international organizations.

The plant breeder must be responsible for initial increases when the program is in development stages 1 and 2. But, by the time the program reaches stages 3 or 4, maintenance and multiplication become a burden for the breeder and may be shifted to a department within the same research establishment or to a separate organization (Basic Seed enterprise). A separate organization may be able to handle maintenance and multiplication for several departments. The last responsibility a plant breeder should be relieved of is variety maintenance. And even when a Basic Seed organization is formed, the breeder must remain a working partner. The personnel to whom this work is entrusted require special skills and training.

In planning initial seed increases, careful thought must be given to how each variety is to be maintained, the need for storage space, the number of multiplication cycles required, and the amount of seed needed to support later multiplications. The facilities needed include a seed processing plant to handle both small seed lots and larger quantities of Basic Seed, plus suitable space for long-term storage to avoid having to produce every stage of maintenance and multiplication every year. Land must be available for variety maintenance and the initial multiplications to Basic Seed, but Basic Seed production may be contracted to selected farmer–seed growers, especially after a variety has been introduced. (In a seed certification program, Certified Seed is produced from Basic Seed. The generation before Basic Seed is called Breeder Seed. Seed that is marketed without certification is called commercial seed and may have been grown from Basic Seed or an equivalent seed stock.)

Seed from the initial multiplications will be used in different ways, depending on the stage of the program. In stages 1 and 2 much of the seed, especially cereal seed, can move directly to farmers for multiplication. In stages 3 and 4 more seed is needed, therefore more multiplications are necessary; consequently, the seed from initial multiplications is used only for further multiplication by seed growers and enterprises.

The way Basic Seed (or its equivalent in programs that do not certify seed) is allocated can enhance or hinder the development and growth of seed enterprises that rely upon public-sector plant breeding. This seed usually should be priced higher than commercial seed or Certified Seed.

The planning of initial seed multiplications requires the combined efforts of administrators, leaders of crop research programs, key personnel involved in the work, and people in seed enterprises responsible for subsequent multiplications of seed.

Building the Seed Supply (Chapter 4)

Importing is often the quickest way to make good seed widely available. Especially for the short term, importing has advantages for supplying seed of minor crops or seed that may be uneconomical to produce locally. The drawbacks of importing are the danger of becoming too dependent on imported seed or a possible hampering of the growth of a local commercial seed industry. But total restrictions against imported seed can work to the disadvantage of a country's seed program and seed enterprises. Finding ways to strengthen the local seed industry makes more sense than imposing prohibitions on imported seed.

Local capacity for seed production should be expanded through seed enterprises. Seed enterprises can be family operations, partnerships, cooperatives, companies, or corporations. The appropriate organizational pattern depends on the scope of the activity, the needs of the enterprise, and the functions to be carried out.

Alternative approaches for developing seed production capacity through seed enterprises include (1) private seed enterprises that do their own plant breeding, production, and marketing; (2) private enterprises that benefit from conventional government seed activities; (3) private enterprises that receive heavy government assistance, but not direct investment; (4) private enterprises that have direct government investment and participation; and (5) government enterprises plus government operations in some or all seed production and marketing activities. In most countries more than one of these types of enterprises exist.

Well-managed seed enterprises can ensure the production and marketing of an increasing quantity of good seed of improved varieties.

Although management of a seed enterprise is similar to management of other organizations, it is complicated by the seasonal nature of the work, the exacting timing and sequence of activities, the dispersal of production over large areas with many seed growers, and the biological nature of seed, which demands conditions that keep it viable. The production process, too, requires planning, the careful selection of areas suitable for seed growing, and finding seed growers who use good agronomic practices. Seed drying, processing, and storage require skilled employees and a reliable quality control program. The financial demands are unique.

Seed production capacity can be increased by (1) taking steps to raise seed yields per unit area, (2) increasing the efficiency and capacity of seed processing equipment, (3) enlarging investment in the seed industry, (4) stimulating the growth of seed enterprises (because profits are possible), and (5) training personnel to improve their managerial and technical skills.

Foreign seed enterprises that have experienced personnel and a wide range of germplasm—especially of maize, sorghum, and vegetables—can help a country's seed program through distributorships, franchise arrangements, consulting services, and various levels of equity participation with local seed enterprises.

Seed Quality Control (Chapter 5)

If a seed program is to succeed, the seed of improved varieties must consistently be better than the seed the farmer produces himself. Concern for quality cannot be left to one person or one organization; it must pervade every aspect of the seed program. Seed enterprises that emphasize seed quality in their own activities are the first line of defense against bad seed.

To help ensure better quality, governments usually utilize seed testing, certification, and legislation. These government quality control measures can be introduced at various times, but the most common sequence is

- testing in stages 1 or 2,
- certification in stage 3,
- general legislation on seed marketing in stage 4.

The emphasis in seed testing is usually on physical purity, germination capacity, and moisture content. Seed health tests, varietal purity evaluations, and seedling vigor tests are sometimes conducted if skilled personnel and suitable facilities are available. Testing results are used in seed certification and seed law enforcement as well as by seed enterprises and farmers; therefore, the seed testing laboratory must be integrated into all of these activities. At least the main seed testing laboratory should have authority to undertake practical research to solve technical seed quality problems encountered from producing to marketing. The International Seed Testing Association has rules and guidelines for testing seed sold internationally, but, with some adaptation, they are also used in many countries to test domestically marketed seed.

There is a rationale for seed certification only when there are functioning seed growers and seed enterprises that use it. A seed certification program involves (1) determining eligibility of varieties, (2) verification of the seed source, (3) field inspection, (4) sampling processed seed, (5) seed testing and evaluation against quality standards, (6) labeling, (7) conducting variety control plots, and (8) education and information.

Seed certification in newer programs helps ensure "trueness-to-variety" and a satisfactory quality for a portion of the seed available in a country. Trueness-to-variety, however, does not imply extreme uniformity. Rather, it means there is good evidence of stability in a variety's composition and performance. Varieties are certified true to the characteristics (including variations) as described by the breeder.

Seed legislation must be kept in proper perspective: it does not create seed. Legislation has little value until seed is being produced and marketed. Good judgment is required to determine when seed legislation is warranted and then to adopt no more legislation than is necessary. Legislation should be designed to increase concern for seed quality, to stabilize quality standards and procedures at practical levels, and to facilitate the growth of seed enterprises and marketing groups. In particular, seed legislation can be adopted to establish a crop research and evaluation system, a seed certification program, marketing requirements for different categories of seed, seed testing responsibilities, a plant variety protection or breeders' rights system, and a plant quarantine program.

For organizational efficiency, it is desirable to concentrate all quality control activities at a national seed center and use a national seed board for guidance on policies. In large programs, state or provincial seed centers linked by a national coordinating mechanism may be appropriate.

Because of the responsibility involved and the skills required, the leaders and technical staff members of quality control activities need extensive knowledge of seed production and technology, as well as the ability to motivate subordinates and to work well with seed growers and leaders of seed enterprises. Personnel should be carefully selected and enabled to remain in the program so they can develop into professional seed technologists.

The success of quality control activities is measured by the quality of seed that consumers get from government agencies, seed enterprises, and marketing groups.

Getting Seed of Improved Varieties Used (Chapter 6)

Until farmers harvest a crop from the seed of improved varieties, no one benefits from the time and money invested in making available good quality seed of improved, high-yielding varieties. Many seed programs emphasize producing and processing seed but neglect factors that contribute to seed use. Planners and managers, both public and private, can strengthen the total seed effort by increased attention to getting seed used.

Leaders of seed activities must be aware of the factors that influence a farmer's acceptance and use of improved varieties. They must establish ways to inform and teach farmers and others about seed of improved varieties. They must encourage the development of a sound seed-marketing system to provide farmers with improved varieties. And they must relate all these points to government policy and actions to help get more good quality seed of improved varieties used.

Availability of production inputs and access to markets for a harvested crop strongly influence acceptance of innovations such as seed of improved varieties. Although many factors affect the rate at which seed of a new variety is adopted, studies show that farmers in some developing countries have adopted new varieties faster than farmers in more developed countries. Campaigns to introduce new varieties and related technology should establish specific goals, identify the groups with whom communication is needed, and commit resources to stimulate action by the groups identified to achieve the goals. In successful past campaigns, leaders have tested materials to be used, identified community leaders and developed their support, provided firsthand experiences, and used various channels to transmit information.

Research activities need to be linked to extension programs to effectively transfer technology to farmers. Applied or adaptive research in farmers' fields is being used more and more by national and international institutions and seed enterprises. This technique makes it possible for farmers, for whom the new technology is intended, to participate, to learn, and to be stimulated to obtain seed of promising varieties. In addition, field agronomists or extension workers learn how to manage the new technology effectively.

The professional staff members of the research programs and extension services who work with farmers need technical knowledge, scientific know-how, economic understanding, farming ability, and communication skills. Successful campaigns require adequate resources for staff training; a budget for salaries, operational and maintenance costs, and travel costs; and personnel policies and incentives to encourage the staff. To be most effective, research, extension, and communication personnel as well as personnel with seed enterprises and marketing groups should cooperate. Informing, persuading, and teaching cannot be considered the monopoly of any one group.

Getting seed used also involves seed marketing. Seed marketing is a tool to foster widespread use of improved varieties. It must be recognized as a distinct activity and different—in organizational structure, method of operation, and staff requirements—from programs designed to inform and teach farmers. Marketing organizations as well as seed enterprises that have marketing programs can successfully assume this responsibility.

Market research at both the national and the seed enterprise levels is necessary to determine the actual demand for seed. Information from market research is needed to organize production and marketing programs. A seed enterprise can forecast market demand by what people (buyers, sellers, or experts) say, what they do, and what they have done.

Market communications include the development of promotional materials, the creation of a favorable impression of the seed supplying organization through public relations, the use of effective techniques to sell seed, and the proper selection and use of dealers for selling seed.

Several marketing channels convey seed from the seed grower to the consumer: the seed grower–seller, the accumulator-wholesaler, the intermediate wholesaler, and the retail dealer. The most appropriate channel depends on such factors as the quantity, value, and characteristics of the seed; the distance from the producing to the consuming area; the financial burden involved; and the amount of service needed before and after the sale.

Procedures for pricing seed may differ considerably from crop to crop, depending largely upon whether the farmer can save his own seed or whether environmental or other factors make this impractical (as it is for hybrid crops, forage species, vegetables, and flowers). Prices should take into account direct costs, indirect costs, profit, and an estimate of what the buyer will pay. The economics of moving seed from the place of production to the place of use requires close scrutiny in planning marketing activities.

Government actions profoundly influence the use of seed. For example, if government policy permits seed prices to reflect all costs as well as some opportunity for profit, the formation and growth of seed enterprises and marketing groups are likely to be stimulated. Seed movement can be facilitated by ensuring adequate credit not only for farmers but also for seed enterprises and other marketing groups. Encouraging well-planned crop production campaigns, by providing adequate support, can stimulate the demand for seed of improved varieties.

Personnel Development and Staffing (Chapter 7)

A country that wishes to improve its seed supplies must make a serious commitment to the assignment of personnel to seed activities, to leadership development, and to the training of personnel. The implementation of programs must be scaled to the trained manpower available. A seed program's stage of development dictates the kind and number of personnel needed. Staff positions can be categorized into decision-making levels and then classified according to various job requirements and organizational needs.

Since seed technology is new in many countries, appropriate positions for the seed technologist often need to be established to give seed technology and the seed technologist a proper status in the agricultural development strategy. In new programs the seed technologist may have to be responsible for several parts of the seed program in order to be able to work on seed full time, but this is better than having several generalists who only incidentally have some responsibility for seed.

Good leaders are needed to develop and maintain a well-trained and qualified staff. Potential leaders should have ability, a good personality, high levels of interest and motivation, creativity, and a willingness to delegate responsibility. Leadership development programs should give individuals opportunities to learn personnel management techniques that lead to improved staff morale, higher motivation, longer tenure of service, increased professionalism, and better performance.

People at all levels often require additional training to be effective seed technologists. Training may be academic or nonacademic; it may be within the country, within the region, or elsewhere. Capable and experienced people to do the training, training materials in a language familiar to the trainees, and opportunities for practical experience are important aspects of effective seed production and technology training.

Resources (Chapter 8)

In developing countries, many vital programs compete for scarce physical, human, financial, and external resources. Leaders of seed programs must obtain adequate resources and use them efficiently to supply

Schematic pattern of the growth of various components of a seed program. All components need not exist at all stages. Growth occurs at different times and at varying rates. Moreover, the needs of small and large programs are different. These illustrations summarize points made about growth throughout this book and may help identify priorities at various stages of development and clarify the growth of components in relation to one another.

CROP RESEARCH	Stage 1	Stage 2	Stage 3	Stage 4
Performance testing	*	* *	* * * * *	* * * * * * *
Plant breeding with: performance testing / stability testing		* *	* * * * *	* * * * * * * *
Regional and international cooperation	*	* * * *	* * * * * * *	* * * * * *
Private breeding			* *	* * * *
Variety release and review committee		* *	* * *	* * *

INITIAL SEED INCREASE	Stage 1	Stage 2	Stage 3	Stage 4
Production by plant breeder	* *	* *	* *	* *
Production by plant breeder plus specialized group		* * *	* * * *	
Production by plant breeder plus separate basic seed enterprise			* *	* * * * * * * * *
Production by seed enterprise with own research and seed increase			* *	* * * *
Seed supplied to farmers	*	* *	*	
Seed supplied to seed growers		* *	* *	* *
Seed supplied to seed enterprises		* *	* * *	* * * * * *

BUILDING THE SEED SUPPLY	Stage 1	Stage 2	Stage 3	Stage 4
Seed importation	*	**	**	***
Production by research station, public sector farms and other seed growers	*	***	**	**
Production by seed enterprises with seed grower contracts		**	****	********

QUALITY CONTROL	Stage 1	Stage 2	Stage 3	Stage 4
Government administrative orders	**	*		
Seed certification		*	***	******
Seed testing		**	****	********
Legislation: marketing control				***
Legislation: seed testing			***	***
Legislation: seed certification			***	***
Legislation: plant variety protection				*
Legislation: plant quarantine		*	***	***

GETTING SEED USED	Stage 1	Stage 2	Stage 3	Stage 4
Education and information	*	**	****	********
Seed spread from farmer to farmer	*	**	**	**
Seed marketing through seed enterprises and marketing groups		**	****	********

a maximum quantity of good quality seed.

The alternatives reviewed earlier offer wide choices for the use of public resources. By concentrating on testing instead of breeding, resources may be saved. Importing seed allows investment in a local seed industry to be reduced or delayed. Public resources can be conserved by encouraging private investment in some aspects of a seed program.

Expensive seed processing plants and impressive seed testing laboratories do not in themselves lead to more and better seed. An examination of the operations of existing facilities may reveal ways to make more efficient use of them. When buildings and equipment are purchased, the special requirements of *seed* need to be kept in mind. For example, harvesting equipment must be easy to clean, and drying installations must be designed for seed, not for grain. The kind and amount of equipment needed for seed processing depend upon the kind and quantity of seed involved, the kind and quantity of contaminants present, and the level of quality standards to be met. In general, seed processing facilities should be adequate to meet essential objectives but should remain as simple as possible.

Developing good storage facilities for seed, especially in the tropics and subtropics, can be one of the wisest uses of resources. Such facilities should be developed early in a seed program. For quality control, vehicles to transport the individuals involved in seed certification and seed law enforcement are the most important equipment. The layout of seed testing facilities should be planned to permit expansion as a quality control program grows.

Equipment for the seed industry need not be highly sophisticated. Some of it may be available or fabricated locally. However, if it cannot be obtained locally, the rather modest amounts needed to buy the equipment abroad should be provided. Labor cannot substitute satisfactorily for equipment in many seed operations.

Proper maintenance of equipment can save huge amounts of money. Spare parts and maintenance manuals should be obtained when equipment is ordered. Funds need to be regularly budgeted for maintenance and purchase of additional spare parts. A staff should be specially trained to operate *and maintain* equipment.

The investment in physical resources should be balanced with the development and effective use of human resources. To keep a seed program growing despite budgetary restraints or to reduce the drain on public funds, administrators should consider how to mobilize private investment and how to make some activities partially or totally self-sufficient. The financial requirements to operate programs and the special credit needs of the commercial seed industry must also be met.

Having funds available at the right time is vital to the operation of a seed program.

External resources can help a country achieve certain objectives quickly. Many bilateral and multilateral organizations offer consultants, grants, training fellowships, and loans. When foreign experts are called on for technical assistance, their responsibilities should be clearly identified, their work should be undertaken in close collaboration with the local staff, clear plans or project outlines should be jointly developed, adequate local support should be provided, and progress should be reviewed periodically.

ORGANIZATIONAL POINTS

A seed program's stage of development, the amount and kind of government involvement, the general organizational pattern of a country's government, and tradition may affect the organizational structure used for a seed program. Methods of organizing crop research, initial seed increases, seed enterprises, quality control programs, educational and informational programs, and seed marketing are discussed in detail in subsequent chapters. This section attempts to relate these components to one another.

Chapter 1 suggests the formation of a national seed board, a seed advisory committee, or some such group to serve as a coordinating mechanism among the various components of the program. Chapter 5 also refers to the role of such a group and, in addition, suggests the formation of a national seed center to link certain quality control functions. Chapter 4 discusses development activities that could be undertaken to stimulate seed production. If this is done, such work needs central direction, a clear focus, and some organizational links to other parts of the seed program. Chapter 1 discusses the planning and Chapter 7 the personnel needed to achieve this objective. Stimulating seed production also should be an integral part of programs as they reach stages 3 and 4.

In mature programs and new ones, the keys to success are having a clear direction, maintaining continuity, concentrating available resources, and allowing the program to grow systematically. The organizational structure can reinforce all of these objectives. Although a total structure is not needed in the beginning, guidelines for the future organizational development of a program should be established. A rational structure will evolve by the time a program reaches stage 4. In the two organizational diagrams shown here, a national seed board and a national seed center are incorporated. Each country, as it assesses its existing structures and needs, can create its own system. The diagrams,

Organizational pattern for a seed program of modest size

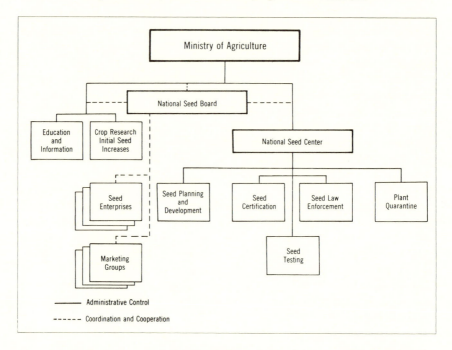

Organizational pattern for a larger, more advanced seed program

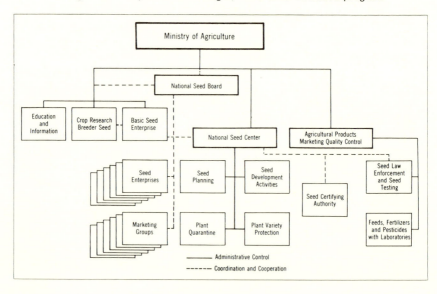

however, serve to identify the functional parts that may be needed as a program develops. In small programs some functions may be consolidated. A useful activity of the seed review team would be to suggest an appropriate organizational pattern for the current program and for the future.

Thus, seed production *today* should not be the only objective in a seed program. In each of the segments of a program, new institutions should be built. Stronger research capabilities, effective ways to maintain and multiply initial seed increases, seed enterprises, quality control systems, marketing mechanisms, and educational units are these new institutions. Success ultimately is measured by whether the institutions are innovative and can support growth in agricultural production over the long run. When this is achieved, regular seed supplies of new, improved varieties will flow to farmers through an ever-enlarging pipeline. The end result, and the final goal, are increased agricultural production and improved well-being.

1
Seed Supplies Today and Tomorrow

Seed is essential to the survival of mankind. Seed stores the highest genetic potential that science can develop and is a vital ingredient in modern agriculture. Seed is programmed, like computers, to hold and give back specific characteristics. Seed of improved varieties permits a farmer to produce a larger crop with more of the qualities he desires.

Agricultural leaders are no longer content with traditional methods. They know that the magic of seed can breathe new life into agricultural programs—they have seen it happen or have read reports of successes in numerous countries. They seek ways to ensure better seed supplies, not only today but also tomorrow. They frequently ask such questions as

> Since most farmers save their own seed, can we really improve our seed situation?
> How important is "improved seed" to agricultural development?
> How can the status of the present seed situation be judged?
> What can be done to improve the seed supply position?

These questions, and many others, harry busy administrators and leaders. Simple answers and "recipes" guaranteed to succeed in all situations do not exist. However, this book can help the diligent administrator who is willing to search for and assemble the building blocks necessary to develop a strong seed program and the resulting seed industry.

Most successful seed programs have developed from small beginnings. Seed program development is analogous to a child's having to learn to creep before he can walk. The development stages must occur in a well-planned and logical sequence. Above all, goals that are realistic in relation to the true needs of the area, and are compatible with ecological conditions and the available or potential physical resources, must be established and maintained.

There are four major steps in moving from today's level of seed sup-

19

Terms Frequently Used in This Book

Basic Seed. A class of seed in a seed certification program that is the last step in the initial seed multiplications and is intended for the production of Certified Seed

Breeder Seed. A class of seed in a seed certification program that is produced under the supervision of the plant breeder, originator, or owner of the variety; is controlled by that person or institution; and is the source of the initial and recurring increases of Basic Seed

Certified Seed. A class of seed that has been certified to conform to the standards for genetic purity established and enforced by a seed certifying authority; is the direct descent of Breeder, Basic, or Certified Seed; and is intended for the production of Certified Seed or for purposes other than seed production

Commercial seed. Seed that is intended for crop production and has not been produced under a seed certification program

Seed. Parts of agricultural, silvicultural, and horticultural plants used for sowing or planting purposes and contrasted to grain, for example, which is used for consumption by humans and animals

Seed enterprise. Any organization involved in seed growing, either directly or through contracts with others; drying; processing; storage; and marketing

Seed grower. One who grows seed but does not process it or sell it to farmers

Seed industry. The entire complex of organizations, institutions, and individuals associated with the seed program of a country; the *commercial* seed industry includes those individuals, seed enterprises, and marketing groups involved in producing and marketing seed for sale to consumers

Seed program. The measures to be implemented and activities being carried out in a country to achieve the timely production and supply of seed of prescribed quality in the quantities needed

(See Glossary for these and other definitions)

plies to a new level tomorrow: (1) decide, and clearly emphasize through policies, that the seed supply is important; (2) determine how seed can be used to support agricultural development objectives; (3) develop a mechanism for examining the current seed situation and future needs; and (4) devise a means to achieve the goals that have been set.

IMPORTANCE OF THE SEED SUPPLY PIPELINE

At harvesttime many farmers save some seed for planting in the next season. They can do this more easily with some crops than with others. In a traditional agricultural system the farmer's own seed is sufficient. However, when attempts are made to introduce a new technology—a new variety, an increased use of fertilizer, a greater degree of mechanization, a different water management regime, or a new cropping system—seed supply channels become critical. If the variety or quality of the seed used is inadequate, expenditures on other technological changes may be wasted. A seed supply pipeline is required to bring good seed to the farmer. Moreover, in order to maintain a change and expand it to new production areas, the supply of seed must be sufficient to meet the demand of innovative farmers. If hybrids are used, the entire seed supply must be replaced each season.

Sometimes when farmers have changed to a new variety, no replacement seed is available so the original material gradually becomes mixed and unidentifiable. If even modest quantities of replacement seed are made available each year, leading farmers will buy and plant it and subsequently sell seed to their neighbors. Thus, change can permeate the entire system. As complex multicropping and intercropping systems and precision planting are introduced, reliable seed supplies become even more important.

Farmers are often more willing to change to seed of an improved variety than to adopt other technological changes. Thus, seed can trigger change and be the vehicle for introducing other changes. Seed of improved varieties can be used time after time to boost production in one region after another, and one crop after another. For this reason, the base upon which continuing seed supplies are built should be solid and capable of growing and changing. Sometimes a crash program or a massive importation of seed has triggered a seed activity. But these devices should be recognized for what they are—measures to meet short-term needs. An administrator concerned with a country's long-term seed needs must look further.

Returns on a government's investment in a seed industry can be high when the system works well. And when a government develops ways to stimulate private investment in the seed industry, the return achieved with the public funds invested can be multiplied. Private seed enterprises can contribute to the national economy through the taxes they pay and through employment of people.

Seed program development requires sustained priority and attention if long-term goals are to be achieved. The seed supply pipeline must be recognized as important at the policy-making level when development plans are made and implemented.

SEED IN SUPPORT OF AGRICULTURAL DEVELOPMENT

The agricultural strategy of most countries involves rapidly increasing production and yields of crops. Numerous factors affect the success of that strategy, including the availability of good quality seed of improved varieties. Seed of improved varieties can support development in both good farming areas and poor farming areas. However, new varieties and new seed supply systems may be required.

High-yielding variety programs have had the most impact in productive areas. Farmers in those areas are able to go long distances to buy seed of the improved varieties and also are able to pay cash or obtain credit. The opportunities for developing effective seed enterprises are good under such situations.

With less favorable production conditions, different crops and other varieties may be needed. For example, in areas of low rainfall, it may be better to concentrate on improved barley varieties than to try to use high-yielding wheat varieties that work in areas with a more suitable climate. In less productive areas, farmers face more risk and may be less willing to invest in seeds and other inputs. Until they feel confident that a new plan will increase production, they may need easier availability of credit and small quantities of seed for trial and multiplication. Special distribution and marketing efforts may be necessary to encourage farmers in these areas to try the seed.

For both favorable and less favorable production conditions, the most effective use of seed to achieve production objectives must be determined. When an adequate seed supply exists, cropping patterns can become more flexible. This is relevant not only in multicropping and intercropping programs but also in regions where precipitation is unreliable and crop failure frequent. The availability of seed of early maturing varieties or of alternative crops can enable farmers to adjust to unexpected and changing situations.

Seed is not just another input. It is a dynamic instrument for change, and it can be used to achieve specific agricultural production objectives.

The question is not what will happen without a good seed program but, rather, how can a good seed program help achieve production objectives?

EVALUATION AND DEVELOPING RECOMMENDATIONS

Having decided that seed of improved varieties is important and that it can promote agricultural development, the next step is to examine the nation's seed situation and to develop recommendations for the future. Can the seed program play a role in the agricultural development strategy and contribute to economic development? What changes are needed? What steps should be taken to improve the program? To answer these questions, seed production, imports, exports, and supply should be reviewed crop by crop. The review can be undertaken as a separate project or as part of an overall review of the agricultural sector.

Four Stages of Seed Activities

The scope of the study, and therefore the amount of effort required, depends on the level of development of a country's agriculture and seed-related activities. Four broad stages of seed activities can be delineated.

In *stage 1*, agricultural research and development are ineffective, limited, or just getting under way. Most varieties of basic food crops are traditional, as are production practices. Nearly all farmers save their own seed, but a plant breeding department may be distributing small quantities of improved varieties of some crops.

In *stage 2*, agricultural research and development are under way. Improved varieties of basic food crops are being developed and are beginning to replace traditional varieties. Use of production inputs, such as fertilizer, is limited but improving. A few farmers and seed production units are multiplying seed. The limited quantities of seed available are a constraint upon improvements in crop production.

In *stage 3*, agricultural research and development are well established and productive. High-yielding varieties of basic food crops are rapidly replacing traditional varieties in the most productive areas of the country. Production inputs are widely used, although usually not at the most efficient levels. Many components of a seed program exist, and the supply of seed ranges from fair to adequate. Seed quality may be poor, distribution remains relatively inefficient, and farmers use much less seed than is available for distribution. Some private seed enterprises are being formed.

In *stage 4*, the agricultural sector is well advanced. The national seed policy is reexamined, special attention is given to developing and strengthening commercial seed production and marketing, a seed law is in force, and links are established with related and supporting institutions and groups.

These stages of development and their characteristics are, of course, artificial categories. The production practices and the seed supply situation may be well advanced for some crops but not for others. Industrial or export crops such as cotton may be in stages 3 or 4 while cereal and forage crops remain in stages 1 or 2. Nevertheless, in beginning any review of a seed situation, the stage of development must be considered. There are far too many attempts to establish stage-3 or stage-4 seed programs when a country is only able to handle stage 1 or 2. Such efforts often fail, wasting scarce human, financial, and material resources.

The Seed Review Team

The review of the seed situation in a country can best be made by a group of specialists—a seed review team. The source of a team's appointment, and the party to which each team member is responsible, should be determined by the purposes of the review, the involvement of external assistance agencies or institutions, and the country's level of agricultural development.

A team is normally appointed by an official such as the minister of agriculture or the director of national planning. A seed review team should have five or six members selected for their experience in such areas as crop breeding research, seed production and supply, marketing and input supply, agricultural extension, agricultural credit, and development planning.

Although the review is "internal," the members of the team can be local personnel only, local personnel plus a few consultants from abroad, or predominantly external experts with local personnel acting in a supporting role. The stage of development of a country's seed activities will influence the composition of the team. When the seed activities are in stage 1, and sometimes in stage 2, local capabilities in seed program development, analysis of needs, and planning are often limited. Therefore, outside sources such as technical assistance agencies, international institutions, or consulting firms should be asked to help. In stage 3, and sometimes in stage 2, local expertise is usually adequate. Nevertheless, the viewpoints of one or two outside advisers can be valuable. And, depending upon the background and experience of the team, an opportunity for its local members to visit and study more advanced programs may improve the team's ability to assess the local situation.

After its appointment, the team has to organize and assign tasks. Since differences often exist among crops, the team will need to consider seed program activities for all major crops unless its guidelines limit it to fewer crops. The team can always call upon specialists to help assess the program in relation to a particular crop.

Areas Needing Review

In reviewing national seed production, supply, and use, the team should seek answers to four broad questions. First, can current and planned seed production and supply activities and policies support planned agricultural development? (If the answer is yes, the team can terminate its activities.) Second, what improvements in the seed production and supply program are needed to support planned agricultural development, or to extract the maximum benefit from agricultural research developments in the country, the international research centers, or elsewhere? Third, what resources are available or required to reach an appropriate level of seed production and supply? Fourth, how should the seed program be organized and where should responsibility be placed to best support agricultural development—on which institutions or organizations, in the private or public sector? Answers to these questions can only be obtained by closely examining key agricultural programs and their components.

Agricultural Development Strategy

The nation's agricultural development strategy should be examined to determine the dependence of its components on seed supplies and the provisions made for ensuring needed seed supplies. A major task of the seed review team will be to assess the degree to which seed production and supply support overall agricultural development, or the extent to which inadequacies in seed production and supply impede development. The findings and recommendations of the team must be relevant to the total agricultural development strategy.

Crop Improvement Research

The mission of a seed program is to produce and supply good quality seed of improved crop varieties. Assessments of the productivity, level of support, and proficiency of crop research—especially crop breeding, introduction, and varietal testing—are crucial for determining how a seed program should be improved. The review team should consider not only current crop research but also how crop research can provide a better base for seed program development. Important points to review include

Kinds of crops emphasized in crop research
Number of improved varieties of each crop released in the past
  five years and the type and degree of improvement
Scope and intensity of varietal testing programs
Policies and procedures for varietal release
Policies and procedures for varietal maintenance and production
  of Breeder Seed
Cooperation with regional and international research activities
Opportunities for private and other nongovernmental research

Initial Seed Multiplication

If the crop research programs are releasing improved varieties, the
procedures for getting the seed multiplied and used should be reviewed.
When seed activities are in stages 2, 3, or 4, special staff members or
units are often involved in the first cycles of seed multiplication. Points
to review are

Responsibility for the different cycles of multiplication
Procedures and arrangements for production of the initial generations (these generations are called Breeder Seed and Basic Seed
  in many seed certification programs and in this book); if the initial seed multiplications are not associated directly with crop
  breeding research, the degree of coordination and cooperation
  between crop research and the initial seed multiplication program needs consideration
Kinds (crops), varieties, and quantities of each multiplied as Basic
  Seed or its equivalent
Facilities and personnel
Quality control procedures and experiences
Use made of Basic Seed and policies for allocating Basic Seed

Building the Seed Supply

When seed activities are in stages 1 or 2, there is little seed production
to assess. However, in stages 3 or 4 substantial quantities of seed are
being supplied, and several aspects should be carefully evaluated.

Kinds, varieties, and quantities of seed produced for distribution
  to farmers
Seed production arrangements and potential
Kinds of seed enterprises and opportunities for growth and for
  profit incentives to develop

Advantages or drawbacks of local production of seed
General quality of seed produced; quality control procedures
Kind, quantity, and quality of seed imported and exported

Steps Taken by Government to Improve Seed Quality

Although producers and suppliers of seed must develop their own quality control measures, the government often institutes quality control measures too. By the time a program reaches stages 3 or 4, the benefits from these activities can be great. The team will need to consider:

Quality of seed being supplied to farmers
Current or potential value of a seed certification activity
Nature and effectiveness of quality control activities at the marketing level not only for seeds but also for fertilizers and other inputs
Administrative or legal steps taken by government to improve seed quality

Getting the Seed Used

An educational program is needed to demonstrate to farmers the benefits of using improved seed and cultural practices. Farmers also need to learn how these improvements can fit their cropping patterns. The seed review team must determine whether programs and activities for making seed available and getting it used support the seed production activities and fulfill the objectives for the use of improved varieties. Special attention needs to be given to

Emphasis of the current educational effort and the type and location of demonstrations
Quality of communication capabilities including mobility of the staff
Credibility of the extension program among farmers, agricultural businesses, and researchers
Involvement of the educational and field staffs in input supply and in distribution, agricultural credit, and marketing
Accuracy of demand projections for seed needs and uses
Distribution and marketing arrangements, including seed pricing structure and policies
Methods farmers use to acquire seed of improved varieties other than through an organized seed program
Extent to which improved varieties have replaced traditional

varieties by kind, variety, farm size, and geographical or political subdivision

Level of coordination and cooperation between the extension program and programs in crop research, agricultural credit, and
input supply

Resources Available and Required

An appraisal should be made of the physical, human, financial, and
external resources available and required. Points to consider include

Bulk storage, drying, cleaning, packaging, and flat storage
Availability of transportation—for the seed and for staff members
Capacity and use made of the seed testing facilities
Number and level of trained personnel; need for further training
Amount spent on each component of the seed program
Internal and external sources to meet future financial needs
Possible need for foreign assistance

Supply of Inputs Other Than Seed and Credit

The full benefits from use of improved seed are generally achieved
when other improved crop production inputs and techniques are introduced too. Along with improved seed, fertilizers, water, pesticides,
herbicides, implements for better land preparation, etc., must be
available to the farmer at a price he can afford. Input availability,
distribution procedures, and the agricultural credit situation should be
examined by the seed review team to determine their adequacy for current and expanded seed campaigns. Points to assess are

Availability and prices of fertilizers, pesticides, herbicides, tillage
implements, irrigation supplies, and other inputs
Availability and terms of financing to farmers
Public and private involvement in supply of inputs
Distribution and marketing procedures
Level of farmers' use of inputs by geographical or political region
Involvement of input suppliers and distributors in seed produc-
tion and supply (or their interest in participating in seed pro-
duction and supply)

Development and Agricultural Policies

Development and agricultural policies can limit seed production, supply, and use as severely as technical, financial, and personnel deficiencies. Agricultural policy is usually beyond a seed review team's in-

fluence, yet the team must examine the country's agricultural policy to determine how it affects the seed industry. Useful areas to examine include

> Policies on prices, subsidies, and imports of agricultural equipment and inputs and their effect on the seed program
>
> Restrictions on movement of agricultural commodities and the extent to which "seed" is treated as "grain" or another commodity instead of as an input
>
> The opportunity for businesses to participate in agricultural development, especially in production and supply of improved seed
>
> Restrictions on import and export of seed, including experimental quantities needed for testing and material for multiplication
>
> Policies on foreign participation in input enterprises

Coordination and Communication

In assessing a seed program and related activities, coordination among the various components must be considered. Duplication, waste, and lack of communication weaken programs and retard growth. A review team can identify these problems and suggest changes.

The foregoing discussion of the various components, activities, and factors to be assessed by a seed review team is not all-inclusive. During preparation for a review, each member should acquaint himself with later chapters in this book and the material in *Seed Program Development* by Delouche and Potts, *The Role of Seed Science and Technology in Agricultural Development*, edited by Feistritzer and Redl, and the Food and Agricultural Organization's (FAO) *Improved Seed Production*, edited by Feistritzer and Kelly. This background on crop production policy, analysis of seed status, requisites for a seed program and industry, relations among components of the program and with related activities, formulation and preparation of seed program plans and projects, and the role of improved seed in agricultural development will suggest other points to assess. A review team will probably develop its own list of specific points to consider as it studies a program.

Improved seed is so integral to agricultural development that it may be tempting to expand the seed status inquiry to encompass the entire crop aspect of a nation's agricultural sector. A team must keep seed program development in perspective and not attempt too much. A seed program and the resulting industry compose one important component of an agricultural development strategy. This component is neither an end in itself nor a panacea for the difficulties that may beset the entire agricultural development effort.

Identifying Needs and Developing Recommendations

After a seed review team has reviewed the seed situation and established the importance of each aspect, it should be able to give definite answers to the four broad questions posed earlier. The analytical process for developing answers involves collating, organizing, and interpreting facts from several viewpoints.

Stage of Development

The stage of development of agriculture and the rural infrastructure of a country determine what type of seed program is economically feasible. This elementary fact is often overlooked. For example, in some countries, seed testing laboratories have been established, and allowed to deteriorate, years before the country's seed production and supply became significant enough to use them. Elaborate seed laws and regulations have been proposed a generation before substantial commercial seed production and marketing occurred. Modern seed processing facilities have been constructed far from electric power and all-weather roads. Proposed improvements in seed programs must be in harmony with the development momentum in agriculture, and seed program objectives ought to be of the same scale as overall agricultural development objectives.

Scope of Seed Production and Supply Operations

In countries where seed activities are in stages 1 or 2, agricultural development is often concentrated on a few crops or in a few regions where substantial improvements appear possible. Development of a seed program should be similarly concentrated. It should support the main development effort and not disperse the limited resources on other crops or other areas.

The seed team should recognize that when the seed program is far behind crop improvement research and other activities, the program cannot catch up in one leap. Progress must be made a step at a time, with each step firmly based on the preceding one. A program that struggles to produce and distribute one hundred tons of seed in one year cannot be expected to produce ten thousand tons the next year.

Quantities of Seed Needed

Projections of the quantities of seed needed must be based on replenishment factors that are reasonable for a country's stage of development. The replenishment factor is the percentage of the total amount of seed needed that is to be produced. In stage-1 and usually in stage-2 seed activities, a replenishment factor of 5 percent for basic food crops with improved varieties is a good objective. Seed of an accepted improved crop variety diffuses rapidly throughout the area for which it is adapted. A new, improved variety can replace existing varieties much more rapidly than the replenishment factor seems to imply. If a hybrid crop is involved, the entire seed supply must be replaced every year. Some stage-1 and stage-2 programs have also found it advantageous to replace the seed of highly commercial, industrial, or export crops every season.

In the more advanced stage-2 and stage-3 programs, a reasonable seed replenishment factor for basic, nonhybrid crops would be 5 to 20 percent. Cross-pollinated crops such as maize composites should be replenished frequently while replenishment factors for self-pollinated crops with high seeding rates, such as certain edible grain legumes, can be nearer 5 percent. Even programs that have passed through stage 4 do not necessarily have high replenishment factors for crops such as wheat, soybeans, and rice. In the United States, the factors range from 10 percent to 50 percent depending upon the area and the crop.

Balanced Development

A seed program consists of closely linked and dependent, or at least semidependent, components (Figure 1). Since the program is usually no stronger than the weakest element, development must be properly balanced. The usual operational and service components of a seed program are

OPERATIONAL COMPONENTS

1. Plant breeding, variety assessment, and varietal maintenance
2. Initial multiplications (Breeder and Basic Seed production)
3. Production and processing

Contracting	Cleaning and grading
Seed growing	Treating
Harvesting	Packaging
Transportation	Storage
Drying	

4. Marketing
   Determining needs
   Accumulation of seeds and services
   Communication
   Distribution
5. Quality control (within enterprises and organizations)

SERVICE COMPONENTS

1. Quality control
   Source verification   Seed testing
   Field inspection   Labeling
   Equipment and   Quarantine operations
     facility inspection   (on imported seed)
   Seed sampling and inspection
2. Education and information services (staff, seed grower)
3. Demand forecasts and marketing intelligence

Later chapters deal in detail with administrative and management decisions associated with each of these components. *Seed Program Development* by Delouche and Potts and the FAO's *Improved Seed Production* will also be helpful to team members as they consider what is required to have balanced development.

Many cycles of multiplication must occur before abundant seed supplies are available for farmers. Impurities, mistakes, poor production, overtreatment, bad storage, weak marketing, and innumerable other deficiencies can rupture the supply chain. Attempts to produce Basic Seed without good Breeder Seed supplies, elaborate inspection services with little seed to check, and seed supplies in storage with no marketing system are signs of poorly balanced programs. A seed review team should recommend improvements in individual elements of a seed program to bring the entire program into balance before turning to any expansion in production.

Facilities and Equipment

Seed production and supply operations require land, equipment, transport, buildings, agricultural chemicals, packaging materials, and other items. The investment needed for a large program is substantial.

It is sensible to organize production, drying, processing, and storage facilities into enterprises or units of a manageable size rather than into a single large monopoly. Yet some concentration of activities and facilities simplifies the contacts with seed growers and other production units, the efforts at quality control, and the requirements of transportation. To handle the volume of seed produced and to maintain quality, at least a moderate level of technology and mechanization—threshers, trucks, dryers, conveying and handling devices, cleaners, treaters, and packaging systems—is necessary. Storage facilities must be capable of adequately preserving seed germination levels (see Chapter 8). A seed review team should recommend equipment and facilities compatible with the size of the enterprises, units, or programs proposed.

Figure 1. Components of a seed program

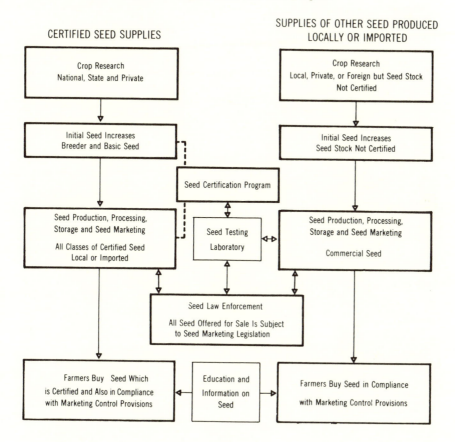

Adequacy of Financing

In stage 2 or later stages, a seed program requires investment capital to purchase land, equipment, and facilities, plus working capital to finance seed production and purchases, maintain facilities and equipment, purchase supplies, cover the costs of marketing, and pay wages. As soon as significant amounts of seed are produced and stored for sale, the commercial seed industry needs large amounts of working capital (see Chapter 4).

At least in the initial stages, the government must supply funds for crop research and varietal maintenance, initial seed increases, quality control measures, education, and information. The need for public funds for commercial production and marketing activities, assistance to seed enterprises, and purchases of processing equipment depends upon policy decisions and whether private businesses are becoming involved.

A seed review team should consider how existing seed operations are financed and what levels of support would be needed to expand activities in the public or private sector. Its recommendations should address the critical need for adequate working capital throughout the seed production, storage, and marketing cycles. The team should also estimate how long it will take various components of the seed program to become partially or totally self-supporting (see Chapter 8).

Seed Legislation and Quality Control

Seed legislation and quality control are easily overemphasized. In stage-1 programs and many stage-2 programs, formal seed legislation and control are rarely necessary. Quality control programs are needed, but these can be conducted through normal management procedures if seed production and supply are predominantly in the public sector.

For more advanced stages, the review team should carefully examine existing legislation and control programs and determine what steps would further the development of the seed program. In particular, recommendations for improvements in plant quarantine are often needed (see Chapter 5 under "Drafting Seed Legislation").

Organization

The organization of seed programs varies widely from country to country. The organizational structure itself appears to be less correlated with a seed program's success than are effectiveness and efficiency. When poor organization impedes progress, recommendations should be made to overcome this limitation. The organization of seed enterprises, quality control programs, and other aspects of a seed program are covered in later chapters. A seed review team should find it beneficial to review relevant sections of those chapters before making recommendations on organization.

Management and Personnel

If a seed program's managers and specialized personnel are competent, operations probably will be efficient. Leaders of a seed program and of components of it must be informed, resourceful, motivated, and dedicated. Well-trained supervisors and specialists are essential, and laborers must be trained and adequately supervised.

A seed review team should make recommendations for management training, technical training, and on-the-job training (chapter 7 can be helpful to a team in this matter).

Resources Available and Needed

It is a good practice for the review team to establish needs and make tentative recommendations independent of what resources are available. After developing tentative recommendations on what *should* be done, the team should examine the resources available and determine what *can* be done. In evaluating resources, attention should be given to unused or underused facilities, equipment, personnel, and land within the seed program and other programs. Frequently, excellent buildings constructed for other purposes could be used for seed program operations. Similarly, underused land, personnel, or equipment might be shifted to the seed program. It should be recognized that what is done is frequently a compromise between what needs to be done and what can be done in terms of readily available resources.

For public seed programs, resource availability is ultimately determined by the government. The private sector, too, may need government assistance. The government can make resources available to the private sector, or it can obtain the resources from foreign assistance agencies or international financial institutions, or it might work through foreign seed enterprises. The seed team's review, assessment, and recommendations will be major factors in what decision is made. Any government has to allocate its scarce development resources to benefit the country as a whole. If the seed review team does not make a good case, the seed program will get low priority. If the team makes a good case, the government can allocate the resources needed or use the team's report in seeking an external loan or grant.

ACHIEVING GOALS SET BY THE REVIEW TEAM

A seed review team's report should identify needs and give clear recommendations for meeting them to the administrator who can then establish well-defined objectives for improving the seed program, develop a plan of action, and phase the steps needed to achieve the goals. The final requirements are the means and the will to make the plan work.

Well-Defined Objectives

The objectives for each component of a seed program should be clearly defined before any activity starts. Unclear objectives cause drift and indecision. The review and evaluation of the existing seed situation should have clarified the objectives.

A Good Plan

A final responsibility for a seed review team is to recommend how the

---

Achieving the Goal: More and Better Seed

1. Decide the seed program needs improvement
2. Form a seed review team
3. Review and assess the nation's seed program
4. Identify needs and develop recommendations
5. Prepare a phased plan of action
6. Take steps to achieve the objective for each sub-unit of the plan

---

program can more effectively support the nation's development. With these recommendations, a plan can be developed. The plan should analyze each component of the seed program in relation to the organizations involved or expected to be involved, the various activities that are to be carried out by each organization, and the specific results to be achieved through these actions. The chapters that follow should be useful in analyzing the present position, future needs, and steps to be taken for each component of a program.

The FAO handbook *Improved Seed Production* provides guidance on "Preparation of a Seed Industry Program" and "Formulation of a Seed Industry Plan," primarily from the viewpoint of a donor or an external assistance agency. The emphasis here, however, is on an internal review made to help formulate agricultural development plans, which may or may not relate to an appraisal by a prospective donor.

Scheduling the Plan

All aspects of a plan cannot be carried out at once. As a plan develops, decisions are needed on which parts will be put into effect in the short, medium, and long term. The FAO's *Improved Seed Production* gives several reasons for developing a seed program in phases. First, since the seed program should be an integral part of the national development plan, which is usually set in three-to-five-year time spans, it should be scheduled in phases that will fit into a series of such plans. Second, phasing permits resources to be concentrated on the most urgent needs in a seed program. Third, the establishment of realistic objectives is facilitated. Fourth, a logical and progressive development of a complete and balanced seed program is made easier. Fifth, interest in the seed program is sustained since phasing establishes at least the intention of a long-term continuity of effort.

The most successful programs have recognized that it takes time to accomplish certain objectives. Seed programs have too many parts to be put into operation quickly. Especially in stages 3 and 4, one component often must be operating successfully before satisfactory performance is possible in another. Therefore, the phasing of the development of one component needs to be considered in relation to the development of others.

To develop a phased program, priorities must be established. Which phases and what portion of each phase will be undertaken first? Establishing priorities and phasing a program make it possible to focus the impact of limited resources and manpower.

Ultimately, the overall plan is divided into sub-units to be achieved by one or more groups. These sub-units may be identified as projects, activities, or numbered phases. Formulation, implementation, and evaluation of projects are discussed in *Improved Seed Production* published by the FAO. Details must be well planned if success is to be achieved.

Making the Plan Work

Since a seed program has many parts, the various units must cooperate closely. A special coordinating mechanism such as a national seed board or a seed industry advisory committee can foster cooperation among the many units that make up a seed program. Such a group serves not only

---

Seed Supplies: Major Policy Points

1. Whether seed is or is not important to the nation's agricultural development strategy
2. Whether the present national seed situation does or does not justify a systematic review and assessment
3. Whether a seed review team is to be formed, and, if so, what its composition and its mandate should be
4. Whether to accept or reject the recommendations of the seed review team
5. In accordance with the priorities for the seed program and for each sub-unit, what actions to take to achieve the recommendations
6. If a coordinating mechanism such as a national seed board is to be established, its composition and responsibilities

as a focal point for seed program development, but also provides continuity of purpose and consistency of effort. *Improved Seed Production* suggests that such a group should (1) establish concrete national policies for seed production and supply; (2) develop a seed industry program specifying scope, organizational structure, level and methods of financing, broad operational guidelines, and an implementation time frame; (3) recommend administrative or legal actions to eliminate impediments to the development of the seed program, to foster the participation of the private sector in the program, and to increase the incentives for farmers to use improved seed; (4) make recommendations about the regulation of seed activities; and (5) continuously monitor and guide the program to ensure that it develops properly, that it is coordinated and efficient, and that it contributes to the agricultural development objectives.

Administrators and managers concerned with a seed program will translate a seed review team's recommendations into a plan that can be implemented in phases with proper priorities. The plan will require certain policy decisions. Subsequent chapters discuss alternatives and policies.

REFERENCES

Delouche, J. C., and Potts, H. C. 1971. *Seed Program Development.* Mississippi State: Mississippi State University.

FAO. 1975. *The State of Food and Agriculture 1975.* Rome.

Feistritzer, W. P., and Kelly, A. F., eds. 1978. *Improved Seed Production.* Rome: FAO.

Feistritzer, W. P., and Redl, H., eds. 1975. *The Role of Seed Science and Technology in Agricultural Development.* Rome: FAO.

Kanwar, J. S. 1971. Research for Effective Use of Land and Water Resources. In *National Agricultural Research Systems in Asia,* ed. A. H. Moseman, pp. 214–225. New York: Agricultural Development Council.

2
The Genesis: Crop Breeding Research

A seed program is undergirded by crop breeding research and thrives when new and improved varieties are regularly introduced for multiplication. An examination of crop breeding research and its relationship to the seed program is the first step toward a strong seed program. An administrator needs to know if the country's crop breeding research is developing the right varieties, if genetic resources outside the country are being adequately utilized, if private research is being encouraged, and if ways exist to get new varieties released, recommended, and used promptly.

Administrators and leaders of research programs must address a host of technical, policy, and financial options such as what kind of variety to develop, whether to concentrate on crop breeding or only on testing of varieties, the possibility of using commercial trade channels to introduce varieties and assist in technology transfer, how much emphasis to place on a crop breeding program, what types of research activities to encourage, what policies to adopt on transferring germplasm and importing breeding stock, and how to use various kinds of testing. Decisions on these points can have far-reaching effects on the production and marketing of seed.

THE FOUNDATION: IMPROVED VARIETIES

The success of crop breeding is measured by the end product—the variety. A variety is a subdivision of a species and is composed of a group of plants that are distinct from other groups and populations and are identifiable from generation to generation. A variety may be an open-pollinated individual variety, a synthetic variety, a composite, or a hybrid. (In this book the term *variety* includes multilines and blends of lines—definitions are given in "Guidelines for Classifying Cultivated Plant Populations" in Appendix A.)

Desirable Characteristics

Varieties have certain characteristics that the plant breeder can manipulate. The plant breeder's task is to develop improved varieties that can be identified and that consistently perform better than existing varieties. The characteristics that affect the value of a variety to farmers include high yield, disease and insect resistance, agronomic features, and quality. A new variety's characteristics must be acceptable to farmers or the variety will not contribute to increased agricultural production. The seed grower also must find the variety satisfactory if he is to multiply its seed.

Identity

The features of a variety should make it recognizably different from other varieties. These features may be size, height, or color or a characteristic such as disease resistance. To be readily identified, the features of the plants of a variety should be reasonably uniform. However, the level of uniformity that is expected of varieties in developed countries is often excessive for developing countries. Genetic variability may even be valuable if it extends the variety's range of adaptability. But, variability is undesirable if it reduces the yield or quality or if it creates large maturity differences. If the originator of a variety feels variation is useful, the variety and its normal range of variation should be described when it is introduced. Usually, however, breeders try to develop uniform varieties, which makes maintaining identity and purity much easier.

High Yield

Farmers need varieties that have predictably high yields under the conditions for which they are recommended. No variety can be superior under all conditions. One objective of crop breeding research should be to develop a number of high-yielding varieties for an area. In an area planted to several high-yielding varieties, the range of genetic variation is broader than in an area planted to a single variety so the risks are lower. Yield varies greatly from one area to another or even from one field to another within the same area. The yield of a variety can be determined only when it is grown throughout an area and compared with known varieties grown under similar conditions.

Some varieties are more stable than others: they consistently produce high yields under a wide range of soil and climatic conditions. Several years of yield tests in farmers' fields can reveal not only the yield stability of a variety under different ecological conditions, but also other factors that may affect acceptance of the variety. Farmers want consistent and predictable performance as well as high yield in varieties.

Disease and Insect Resistance

Farmers prefer varieties with resistance to local disease and insect pests. However, some varieties may be grown even if they lack resistance provided they mature early enough to escape attacks by the disease or insect. Varieties that consistently produce good yields usually have some tolerance or resistance to diseases or insects common in the area. Extensive testing can help expose susceptibility in new varieties before they are introduced to farmers.

Agronomic Features

The agronomic characteristics of a variety can be as important as yield for farmers. Early maturity is vital in areas where the growing season is short or where crops are meshed into multiple-cropping systems. Strong root systems, stalk strength, and height can affect a variety's acceptance by farmers because these factors influence a variety's adaptation to higher plant populations and fertilizer rates. For forage crops, quick growth and early recovery after grazing may be desirable to provide increased forage as well as competition with weeds.

Matching the variety's characteristics with the farmer's desires

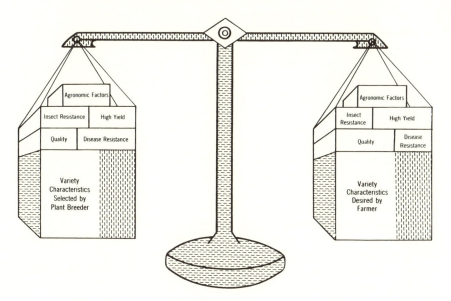

Quality Factors

For some crops, consumer preferences rather than yield or agronomic advantages dictate the choice of a variety. A variety with an undesirable or unfamiliar eating quality should not be multiplied on a large scale until the potential for changing consumer preference has been assessed.

Chemical composition is one quality factor. For example, a higher protein content or a better amino-acid balance in cereals may be nutritionally desirable. If the yield is low, however, farmers are unlikely to plant the nutritionally improved variety unless there is a price advantage to offset the yield loss. In some crops—forages, for example—chemically harmful substances may be present. Crop research programs require access to laboratories that can test for the presence of such substances.

Farmer and Seed Grower Acceptance

The characteristics just discussed are important to a farmer's acceptance of a variety, but when a crop is produced for market, profit is a major concern of the farmer. Chapter 6 deals with other factors that affect a farmer's decision about a variety, and administrators and research workers concerned with varieties proposed for use need to be familiar with all these factors. In recommending varieties to a farmer, it is wise to specify the most desirable agronomic conditions for each variety. The farmer can then select the ones most suitable for his conditions.

For a variety to achieve widespread use, however, it not only must satisfy farmer preferences, it also must be easy to multiply. Some high-yielding varieties fail to become widely used because breeders or seed growers have trouble producing seed. Sometimes multiplication in an environment other than the area in which the seed is to be used can overcome the difficulty.

NATIONAL BREEDING VERSUS TESTING IMPORTED VARIETIES

Administrators frequently need to set priorities in crop research and development. One difficult decision is whether to breed new varieties, with the requisite testing, or to concentrate only on testing varieties introduced from outside the country. In the early stages it usually is best to test introduced varieties, which could be distributed to farmers quickly, because ten to fifteen years may pass between the start of a domestic breeding program and the time when seed of varieties produced by it reaches the farmers. At least six major factors will affect the decision whether to "breed" or to "test":

1. Importance of the crop and need for improved varieties
2. Uniqueness of the growing conditions
3. Future potential of the crop
4. Institutional resources available
5. Human, financial, and physical resources
6. Availability of good varieties and seed supplies

A detailed examination of these points follows because this basic decision is so important to administrators.

Importance of the Crop and Need for Improved Varieties

The economic and political importance of a crop to a country may determine whether a breeding program should be started. A need for the production of seed within the country may also be a consideration. Sometimes a breeding program is warranted in order to develop improved varieties of a crop that is not widely grown but may have potential in the future.

Factors limiting the use of available varieties—such as length of the growing season, drought, salinity, prevalence of certain diseases, scarcity of labor, or waterlogged soil—may make it desirable to breed new varieties especially adapted to these conditions if suitable ones cannot be found through testing introduced varieties. Drainage and irrigation projects, new weed control methods, or better storage could create a demand for new varieties and, therefore, make a breeding program necessary.

In all cases the importance of a crop dictates the need to make new varieties available either through local research or through the introduction of varieties developed elsewhere.

Uniqueness of the Growing Conditions

The country's growing conditions should be well understood. Although excellent varieties of a crop may exist elsewhere, differences in climate, soil, and other agronomic factors may reduce their yield potential locally. Varieties may have to be bred for specific local growing conditions.

Potential of the Crop

Some countries earn large amounts of foreign exchange by exporting certain crops. If the export market for a crop is based upon special quality factors, breeding varieties preferred in world trade can then be justified even though the qualities of those varieties are irrelevant to local preferences.

When the production of a crop begins to exceed local demand, a change of emphasis in quality characteristics may be needed to suit export markets. However, high yield and desirable agronomic characteristics must be maintained. If the needed qualities are not available in local varieties, or if introduced varieties with the needed quality factors do not perform satisfactorily, a breeding program to combine these qualities with local varieties may be desirable.

Survey of Institutional Resources

In determining what priority to give to a breeding or a testing program for a crop, an administrator should survey the strengths and weaknesses of local public and private institutions available to deal with the problem, so that existing resources will be used efficiently. The publication *Building Agricultural Research Systems in the Developing Nations*, by A. H. Moseman, is extremely helpful in making such a survey.

If private seed enterprises have well-developed capabilities and facilities, ways to reduce duplication and maximize results should be explored. Such steps may result in the redirection of some public research to complement private efforts for the mutual benefit of both.

Human, Financial, and Physical Resources

A country's human, financial, and physical resources must be considered in deciding whether to start a breeding program or to concentrate on testing varieties introduced from outside the country. A breeding program requires careful planning, considerable expertise, permanence of personnel, and long-term availability of financial and material resources. If there is uncertainty about any of these factors, the money and time invested in breeding is unlikely to pay off within a reasonable period of time. Under these conditions, government institutions or private seed enterprises should concentrate on testing introduced materials instead of initiating breeding programs.

Availability of Good Varieties and Seed Supplies

A survey of varieties and seed supplies available in the country and in world markets is important when starting a breeding or testing program. Tremendous progress in agriculture has resulted from transferring large quantities of seed of varieties developed in one area to another area. Examples are the massive shipments of wheat seed from Mexico to India, Pakistan, and the Mideast; the movement of seed of improved rice varieties from the Philippines, Colombia, and Surinam to several Central and South American countries; the expansion of sorghum production by introducing sorghum hybrids developed in the United States into Mexico, Central and South America, South Africa, Australia, and southern Europe; the transfer of sugarbeet hybrids from Europe to the United States, South America, the Mideast, and Asia; and the transfer of sugarcane varieties from Indonesia, Hawaii, Puerto Rico, Australia, Peru, and the southern United States to many countries around the world. Both public and private organizations have played a role in testing and transferring varieties.

The climatic area into which a variety is being introduced must be similar to the environment in which the variety was developed. For example, the spring wheat varieties developed in Mexico have been introduced into many parts of the world, but they have been most successful in areas such as India, Pakistan, and the Mideast, which have climates similar to Mexico's. Diseases and insects also differ from one area to another and can limit the transfer of varieties. To ensure success, a testing program should precede any wide-scale introduction of a new variety.

Obtaining Optimum Results—An Analysis of the Alternatives

In deciding between breeding and testing, nontechnical factors, the possible return on investment, the time available, and the support possible from the international research institutes should be considered. Nontechnical factors, such as political implications and the prestige of a breeding program, can override other considerations, but they all need to be carefully balanced. Other examples of nontechnical factors are economic factors, realignment of countries, and internal development factors.

The return on investment in plant breeding in relation to the potential returns to the farmers should be assessed carefully. R. W. Jugenheimer has estimated that the United States receives an annual dividend of seventy-five dollars on each dollar originally spent on hybrid maize research. The cost of testing available varieties versus the breeding of new varieties needs evaluation. Several studies in developing countries indicate that the rate of return on investment in crop research is high, but it is unclear how much resulted from plant breeding within the country as distinguished from the introduction of varieties developed elsewhere.

The time it takes to develop improved varieties is important in deciding whether to concentrate on breeding or testing. Using crop varieties already available may give higher returns and an earlier payoff for a country than will breeding new varieties.

International research centers (Appendix H) develop germplasm and breeding material and make it available in various stages of improvement. Because of the diversity of material available from these centers, some countries concentrate on testing and selecting promising material from segregating lines without initially breeding their own varieties.

USING COMMERCIAL TRADE CHANNELS

Although public crop research through breeding or testing introduced materials probably will be the principal source of new varieties for major crops in most countries, commercial trade channels have introduced and developed new crops, increased production, and contributed to the agricultural technology of many countries. Complementing the local testing activities, commercial seed enterprises can identify and assess needs, locate suitable varieties for testing, and make massive transfers of seed quickly. Private companies were instrumental in transferring hybrid maize from North America to Europe, Australia, and New Zealand; sorghum hybrids from North America into several Central and South American countries; large-grain chick-peas from the United States to Peru; sugarbeet and potato varieties from Europe to the Mideast; and vegetable varieties from North America and Europe to many countries in Latin America, Africa, and Asia.

BREEDING IMPROVED VARIETIES

When a decision has been made to develop a breeding program for a crop, the research goals and general aims of a program should be clearly identified and the amounts and kinds of human, physical, and financial resources needed should be determined. Building an effective breeding program requires long-range planning for the resources needed. Interdisciplinary teams and national research activities should develop links with international programs. The extent of private and public research cooperation needs consideration, with clear guidelines on how the two efforts will relate. The execution of the program needs close attention. Consideration is also needed on how to evaluate results, how to make decisions to achieve progress, and how to move the results of research into production systems and make them available to farmers.

Crop Breeding Goals

The objectives for the breeding program should be laid out clearly. Higher yield may be a primary goal, but farmers also have other priorities. The varietal characteristics farmers need should be carefully reviewed when establishing goals.

The types of farmers who will use the varieties and the farming conditions under which the varieties are to be grown will influence the breeding objectives. If progressive farmers with good land are the primary target, research can concentrate on varieties that will give the highest yields under advanced technology. If the farmers are less progressive, if opportunities are limited by the environment, and if few inputs are to be used, yield stability at less than the maximum level may be a prudent research objective. The crop and the objectives to be achieved will influence the emphasis placed on developing hybrids, synthetic varieties, multiline varieties, or other types of varieties.

The time required to obtain results in a breeding research program, the importance and volume of the potential seed market, and the risks involved in reaching the goals need to be considered. Usually the simplest procedure will give the quickest results in the initial breeding program. Identifying short-, medium-, and long-term research goals is essential. Although some goals can be achieved quickly, others are more difficult and it may be several years before they benefit the farmer. When establishing goals, an administrator needs to consider using existing research results from government, private, regional, and international research programs.

The degree of varietal uniformity needed is related to short- and long-term goals. Where higher yields are a primary goal in the short term, the degree of uniformity may be of secondary importance. However, higher yields and increased uniformity are compatible in long-term goals. The time and effort necessary to develop highly uniform populations must be weighed against the benefits that will be derived.

Although a breeder may focus on increasing crop yields, the need to produce seed economically from the varieties developed cannot be ignored. Thus, identifying ways to increase seed yields is also an important goal.

Allocation of Resources

Administrators are constantly concerned with the allocation of resources. Once a decision has been made to support research on a crop, the basic questions are, how much is needed, how should funds be divided between human and physical resources, and for what time period should the budgeting be based? Answers to these complex questions can be found in *Building Agricultural Research Systems in the Developing Nations*, by A. H. Moseman, which deals extensively with this subject.

Interdisciplinary and Institutional Links

A crop breeder's objective of developing improved varieties with certain agronomic, physiological, and morphological characteristics will be achieved more rapidly if there is close cooperation with specialists in related disciplines. Cooperation with other research programs, including private ones, should be encouraged by an administrator because a free exchange of breeding materials and ideas among research programs benefits all crop breeding research. The international agricultural research institutes are good sources of breeding materials for certain crops and of information helpful to the breeder starting a new program.

Execution

A well-trained, competent staff operating with determination and continuity is required for a successful crop breeding program. Careful record keeping, execution of field trials, and analysis are essential for accurate results. Field work is a weakness in many breeding programs. The systematic handling and evaluation of material in timely, planned steps greatly increases the probability of having an efficient and productive breeding program.

Evaluation of Results

A breeding program should be evaluated regularly. Unproductive projects or completed projects should be phased out or redirected to meet new objectives. An assessment of a project might ask whether the goals are still relevant, whether the results are compatible with the established goals, whether the progress justifies continued effort, whether connections have developed with other programs, and whether changes are required in the level of support. A concurrent evaluation of the achievements of other breeding institutions in the country is needed to ensure that maximum use is made of all available breeding materials.

Disposition of New Varieties

Before varieties are increased, ways to make them available to farmers should be planned. No single method will work in all situations. To benefit the farmer, improved varieties or inbred lines must be released promptly. Many excellent research programs have failed because the results of their efforts did not reach the farmers. The breeders must provide leadership in the increase and distribution of a variety.

SUPPORT FOR CROP BREEDING RESEARCH

Crop breeding research can be supported by a government (directly or through universities), by a regional organization, through international efforts, and by seed enterprises, farmers' associations, or cooperatives.

Government-Supported Research

In most countries some plant breeding research is supported by national or local government. Government-supported research is normally done through research stations or agricultural universities that have a research, and often an extension, component.

Work can be organized in different ways. In some countries, especially those with sharply different ecological zones, each research station develops activities independent of the other stations, and poor coordination among stations often results in duplication of work. An integrated system of research is more efficient; that is, one station leads research on a particular crop, and the other stations assist in testing promising selections derived from the breeding program of the main station. For another crop a different station acts as the main center for breeding work, with the other stations assisting in testing.

Universities with crop research programs sometimes operate as part of the national research system. Some universities, however, have difficulty keeping their research closely related to the national research objectives because professors sometimes place too much emphasis on personal research objectives. Linking universities to national research objectives facilitates training and provides students with practical experience.

Occasionally, a special office at one station coordinates research for each major crop. The "coordinator" brings together all stations and relevant disciplines (such as genetics, agronomy, plant pathology, entomology, and economics) to develop common research objectives, to review results on a yearly basis, and to plan cooperative activities. Coordinated research among disciplines is essential to a well-rounded breeding program.

Other research groups, both public and private, can aid a government-supported research program by cooperating in testing, participating on review and planning boards, and helping with evaluation of the research objectives and results. These groups can also provide facilities and per-

Linking research programs

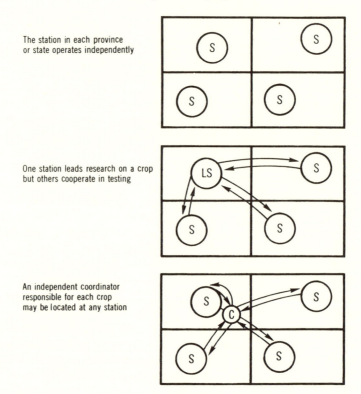

The station in each province
or state operates independently

One station leads research on a crop
but others cooperate in testing

An independent coordinator
responsible for each crop
may be located at any station

sonnel for special evaluations, both chemical and physical, to solve specific problems in a breeding program.

Regional Research

When a crop has potential in more than one country, neighboring countries or countries that lie within the same ecological belt can help each other. Regional research might lead to the development of varieties that have wide ecological adaptation, might permit an exchange of seed for planting, and might foster a more efficient use of the countries' resources. The research activities can be linked through a coordinating committee. Certain phases of the breeding work might be carried out in one or more countries and testing in all of them.

Various kinds of regional cooperation now exist, such as cereal improvement in the Mideast through the efforts of the FAO, Centro Internacional de Mejoramiento de Maíz y Trigo (CIMMYT), the Arab

League, International Centre for Agricultural Research in the Dry Areas (ICARDA), the Ford Foundation, and the Rockefeller Foundation; rice work through the West African Rice Development Association; maize, sorghum, and bean improvement in Central America within the Programa Cooperativo Centro Americano de Mejoramiento de Cultivos Alimenticios; maize improvement in Brazil and the Andean countries of South America through CIMMYT and Centro Internacional de Agricultura Tropical (CIAT); and wheat improvement in East Africa and the Andean region with assistance from CIMMYT.

International Research Agencies

Several international research centers can provide assistance in crop research (Appendix H). Each center not only conducts crop research, but also works with national and regional research institutes. They also train research scientists and production workers. The centers have well-equipped and well-staffed research facilities and libraries and space for conferences, seminars, and workshops. Most of them have extensive collections of germplasm for the crops on which they work.

The centers will provide national research organizations with germplasm for use in research programs, a broad range of plant materials for testing, assistance in training scientists and production workers, guidance on the organization and development of a national research program, literature, and test results. The international centers, however, normally do not maintain seed of named varieties for multiplication and distribution. A nation's own research program must assume this responsibility.

An international network of genetic resource centers is being established under the guidance of the International Board for Plant Genetic Resources (see Appendix H) for the benefit of all countries. A list of locations where germplasm is stored has been published.

Private Research

Private seed enterprises conduct plant breeding so they can sell varieties that are different from those of their competitors. Competition makes enterprises engaged in plant breeding responsive to the needs of farmers. As a result of the research, motivation is high to market varieties developed by the enterprises.

The plant breeding programs of private enterprises take various forms. Some do plant breeding as well as sell seed of the resulting varieties or license the varieties to other enterprises for production. Some concentrate on plant breeding and license or sell the materials developed. Some, rather than breeding, identify and assemble usable genetic material from other sources and produce and market the seed under their own brand names.

In developing countries the amount of plant breeding a private enterprise undertakes will be affected by government policies related to the availability of germplasm, the opportunity to market developed varieties, and seed pricing (other related factors are discussed in Chapter 4). Whether the sale of varieties by someone other than the originator can be prevented may also be a factor (see Chapter 5).

Collaboration between private and government programs has increased through the years in Western Europe and North America and in a few developing countries. Such collaboration takes the form of joint testing, the exchange of some plant materials, and a more distinct division of responsibilities. The results of basic research by public research programs have been made available to all breeders.

The objective should be to use both public and private research capabilities to maximum advantage for the benefit of farmers. Administrators in government are in a key position to help achieve this objective.

POLICIES ON TRANSFERRING GERMPLASM
AND IMPORTING BREEDING STOCK

Germplasm (the source of genetic content or variability transmitted by seed or other plant organs through propagation) is the basic building block for new varieties. No one has a monopoly on germplasm, and all crop research programs benefit when germplasm can be freely exchanged.

Research leaders and administrators often must decide whether to exchange or import breeding material and seed. If clear policies and guidelines are established, the decisions are easier to make and more consistent.

Availability of Seed in Breeding Programs

Policies on the availability of seed from the public research programs should be related to the origin and possible use of the seed. Breeding stocks being maintained without selection in public programs should be freely available to anyone who can make use of them except when seed has been acquired elsewhere and the supplier has placed a restriction on its distribution.

Some breeding stocks may include materials with special characteristics that can be incorporated into commercial varieties. Such material usually should be available to both public and private breeders inside and outside a country. Some breeding stocks may include advanced generations of crosses made to combine desirable characteristics. The plant breeder might adopt a liberal policy on sharing such material with colleagues. Some international research centers have made segregating materials available to other breeders without restrictions.

Experimental varieties are usually not available to others except for testing. After varieties have been released for increase, the Breeder Seed from public programs can also be made available to private breeders. When a hybrid is released, the public benefits if the parents of the hybrid are made available to all breeders so their potential can be used in other combinations. Most public breeding programs in developed countries and in many less developed ones have an unrestricted release policy on such materials. However, in order to maintain a competitive position, private breeders generally restrict distribution of inbreds they develop.

Importing Breeding Stock

If private seed enterprises are being encouraged, policies that facilitate the importation of breeding stocks and seed for multiplication are essential. Private seed enterprises should not be expected to divulge detailed information regarding the pedigree of any material they import or develop. This protection is needed if they are to support their research through the exclusive sale of their own varieties.

Plant quarantine requirements for the flow of experimental materials to and from a country should be as simple as possible. Care is needed to avoid the introduction of insects, plant diseases, and weeds that are not already present in the country. However, tests to detect these contaminants must be made expeditiously and sensibly in order to avoid the loss of valuable time and germplasm (see Chapter 5 under "Drafting Seed Legislation").

TESTING: OBTAINING INFORMATION ON VARIETIES

Both public agencies and private seed enterprises need information about the performance of potential varieties. Materials introduced from abroad as well as locally bred varieties should be tested. The kinds of tests to be performed and the use to be made of the information are related to policy decisions that can have a strong impact on the development of the commercial seed industry.

Responsibility for Testing

The initial responsibility for testing rests with the plant breeder and the plant breeder's organization, whether public or private. The breeder first tests many experimental varieties in a few locations. Varieties that look promising are then tested in more locations. This testing may be done in cooperation with other breeders or institutions.

In addition to the testing carried out by breeders to identify improved varieties, tests are sometimes conducted by governments to assess information on available varieties, to develop a list of recommended varieties, or to determine which varieties are suitable for use. Participation in these tests may be required or submitted to voluntarily.

The testing may be done by research stations or universities. Some countries have a special office that coordinates this work and codifies materials for testing. In other countries independent testing organizations have been established to avoid any possible conflict of interest. This approach has merit, especially for advanced programs that test large amounts of privately developed material.

Kinds of Testing

The appropriate kinds of testing relate to the program's stage of development. Modest performance testing is all that is possible in stage 1. As programs advance to later stages, the performance testing often becomes more intensive and complex. In stages 3 and 4, tests for identity, uniformity, and stability of varieties also are sometimes undertaken if plant variety protection (see Chapter 5 under "Drafting Seed Legislation") becomes a part of the program.

Performance Testing

The most common assessments made by testing varieties include yield, disease and insect-pest reactions, maturity, use characteristics, and standing ability. In advanced programs, these features are evaluated with different fertilizer levels, plant population densities, times of seeding, depth of seeding, timing of irrigation, and other agronomic practices. Therefore, performance testing must be a coordinated effort on the parts of the plant breeder, pathologist, entomologist, and agronomist to ensure the best possible results from the tests.

Performance testing may be done in various ways but three principles should be followed: (1) the geographic and climatic zones in which testing is to be conducted should be precisely defined, (2) the most appropriate varieties and maturity ranges should be allocated to each zone, and (3) within tests, varieties with approximately similar maturities and growth habits should be grouped. After several seasons of testing, response patterns emerge that permit a delineation of ecological districts. Then tested material can be clustered within ecological districts for better data control, and predictions of variety performance will become increasingly accurate.

Administrators must ensure that the test results are dependable. The care with which the tests are laid out, the timeliness of planting and harvesting, the kind of supervision the tests receive, and the appropriateness of the testing locations influence the accuracy of the results.

On-farm testing, by both public and private organizations, begins as promising varieties are identified. Testing in farmers' fields allows the farmer to evaluate the varieties, gives the farmer an opportunity to learn from the testing program, and shows the research and the information specialists how the varieties perform under farm conditions. On-farm testing is more difficult than testing on research stations because the research staff needs vehicles in order to conduct and supervise the tests properly and possibly to transport equipment for planting and harvesting. Standard, locally grown varieties should be included in each on-farm test for comparison with the new varieties.

The mass of data collected from the testing of varieties and the related tests is of primary value to crop researchers as they evaluate progress and plan for the future. Information should be assembled and organized to support decisions recommending the promising varieties for use. Alternatively, information about varieties tested may simply be published. But ultimately, farmers, the seed workers, and those involved in educational work will need the test results on varieties in a clear, concise form.

The administrator or research coordinator plays a critical role in seeing that this last step is done well. Too often research, educational, and seed multiplication activities are poorly connected. Consequently, research results are not put in a form that can be understood by nonspecialists. Some programs have a specialist for the major crops to bridge the gap between the research and information staffs. Seed enterprises and marketing groups play a creative and significant role when they relate their own sales efforts to the general information provided by the research and information specialists (Chapter 6 deals in more detail with these topics).

Testing for Identity, Uniformity, and Stability

The breeder is responsible for properly describing a new variety. Uniformity and stability are important in properly identifying and

describing a variety. In addition, descriptions stating the range of variation expected can be extremely useful.

Information on identity and genetic stability of crop varieties is used
by plant breeders to monitor the purity of varieties, by seed certifying
organizations to assess the genetic purity of varieties, and by seed law enforcement officials to determine whether seed is properly labeled.

When private research and a means for providing plant variety protection exist, testing is sometimes carried out to verify the variety description, to check for uniformity and stability, and to settle disputes. This
kind of testing has not been common in developing countries where
public research has predominated. When plant variety protection is not
involved, a high level of uniformity is less important for identification.
Administrators and leaders in research programs should maintain a practical approach to these issues as programs develop.

Relating Test Results to Variety Release

Test results are used by variety review and release committees, seed
enterprises, regional and international organizations, and farmers for
making decisions, particularly in deciding which varieties to use.

A Variety Review and Release Committee

Plant breeders, in cooperation with pathologists, entomologists,
agronomists, and economists, have the first responsibility to propose
varieties for use from their own research programs. Since many breeding
programs, both public and private, may be involved, a group is needed
to help the government establish policies and to aid in the review, recom-
mendation, and release of varieties. The FAO publication *Improved Seed
Production* states that

> Generally, the decision to release an improved cultivar [variety] is made on
> the basis of a recommendation made by a Cultivar Review and Release
> Committee. . . . The Committee is usually an advisory group of six to eight
> persons representing research, extension, development, and credit agencies
> and the private agricultural community, and appointed by the Minister of
> Agriculture or equivalent. The functions of the committee are to review the
> history and performance record of nominated cultivars, determine their
> potential contribution to the national agriculture, make recommendations
> pertaining to their release and entry into the seed multiplication and pro-
> duction scheme and, equally important, to make recommendations on
> discontinuation of obsolete cultivars.

The committee should be made up of the most qualified persons engaged in varietal development or those who have a specialized interest in seed improvement. Appointment of committee members on solely a political or protocol basis should be avoided if possible.

A variety review and release committee can establish guidelines for considering varieties from public and private plant-breeding programs; determine whether varieties are to be recommended, considered "suitable," or listed as "unsuitable"; establish criteria for accepting varieties as eligible for seed certification; and assume responsibility for an allocation policy for seed of new varieties (Chapter 3, "Policies for the Use of Basic Seed").

For its deliberations, a variety review and release committee needs information and data on the variety and seed supplies: method of identification, performance in comparison with standard varieties, area of adaptation, proposed use, amount of seed available for multiplication or distribution, recommended method of seed increase and maintenance, and proposed method of seed distribution.

Seed Enterprises

A seed enterprise that has a research program also needs a variety release committee, which should include representatives of the enterprise's research, production, and marketing sections. Official testing requirements for varieties from private programs can affect private interest in plant breeding. If private seed enterprises are to be encouraged, policies on the kind and amount of testing expected must be clear. Representatives of the private sector should be invited to participate when guidelines are established.

The growth of crop research program components

CROP RESEARCH	Stage 1	Stage 2	Stage 3	Stage 4
Performance testing	*	* *	* * * * *	* * * * * * *
Plant breeding with: performance testing stability testing		* *	* * * * *	* * * * * * * *
Regional and international cooperation	*	* * * *	* * * * * *	* * * * * *
Private breeding			* *	* * * *
Variety release and review committee		* *	* * *	* * *

Alternatives to official testing are possible. For example, officials might observe the tests made by private plant breeders rather than duplicate them; tests made by an independent agency might be used for official purposes; or seed of any new variety produced by a private enterprise might be required for testing in the official program merely so the results could be published (the use of the variety would not be delayed until such tests were completed—the introducer of the variety presumably would have tested it sufficiently to justify its introduction and use).

Regional and International Cooperation

As regional and international crop research increases, opportunities exist for releasing new varieties cooperatively. If an experimental variety is adapted to conditions in several countries, it could be simultaneously or jointly released under one name by two or more countries. Thus the benefits of cooperation in the early stages of variety development would continue through the release of the variety. Breeder and Basic Seed supplies could be more easily supplied if more than one country were multiplying and maintaining the material, and Certified Seed and other commercial seed of the same variety could be easily traded since the variety would have the same name in all cooperating countries.

A standardization of procedures and materials to be multiplied would be essential in such joint ventures. Since joint release and distribution of a variety would involve two or more governments, policies would have to be established by administrators at high levels to enable breeders and seed production specialists to cooperate across national borders.

The Final Judge—The Farmer

Basic to any decision is the question, how much freedom should the farmer or seed user have to decide what variety to plant? Results from tests and the decisions made by a variety review and release committee could be presented to farmers in several different ways.

1. Provide information about the varieties tested so farmers can choose the ones best suited for their purposes
2. Develop a list of varieties tested, found suitable, and available for use
3. Form a list of varieties not suitable for use
4. Prepare a list of recommended varieties that may be certified
5. Develop a list of recommended varieties that must be certified and are the *only* varieties available for use

The first three possibilities give farmers considerable choice. The fourth provides choice if it is used in combination with at least one of the first three. The fifth possibility limits the choice of varieties, it may restrict seed supplies, and it reduces the economic opportunities of seed enterprises.

In societies where consumers are accustomed to choosing among different products, it is logical to provide information that provides the opportunity for a choice of varieties. Since many factors affect a farmer's

---

Crop Breeding Research: Major Policy Points

1. For each crop, the strategy for research and the relative importance of breeding versus only testing
2. How the crop breeding program will be organized and what level of financial support will be provided
3. Who will be responsible for testing new varieties, what kinds of testing will be needed, what use will be made of test results
4. Role to be played by commercial trade channels in introducing new varieties and crops
5. Whether plant breeding activities by private seed enterprises will be encouraged or discouraged
6. Whether or not breeding stock can be imported by private seed enterprises for testing or multiplication
7. Whether a variety review and release committee is to be established, and what its mandate and composition will be
8. Relation of private plant breeding research (if it is being encouraged) to "official" testing and public plant breeding
9. At what stage and under what conditions germplasm is to be exchanged with public and private plant breeders
10. Nature of agreements and extent of cooperation with regional and international research programs and organizations
11. Nature of agreements with neighboring countries if varieties are to be jointly named, released, and seed stock maintained
12. To what extent farmers are to be free to choose the varieties they want and how much information is to be supplied to them

decision on what variety to use and since some societies limit the choices available to the public, it is understandable that limitations may be placed on varieties. Good decisions made by the state on the choice of varieties may be beneficial to the farmer, but incorrect decisions may harm production or even be disastrous.

Farmers in general are cautious. Their decisions are made with considerable deliberation even when research results show clear advantages for certain varieties. Farmers accept a new variety only after they have good reason to believe that it will perform better for them. Since farmers do not rush in blindly to adopt a new variety, policies on the use of varieties need not be overly protective. Farmers need honest guidance, but they are clearly capable of discriminating among varieties.

Farmers should be encouraged to use the best varieties available. Administrators need to see that policies are clear and that farmers are getting sufficient information to make a wise choice in adopting the improved varieties and related technology that should raise yields and increase economic return. A properly conducted on-farm testing program permits farmers to decide which variety is best for their farming conditions, and such a program merits strong support by administrators. Allowing many decisions to be made by farmers in their own fields alleviates risk and encourages plant breeders to earn the farmers' respect.

REFERENCES

Ardito Barletta, Nicolas. 1971. Costs and Social Benefits of Agricultural Research in Mexico. Ph.D. dissertation, University of Chicago.

Ayer, Harry W., and Schuh, G. Edward. 1972. Social Rates of Return and Other Aspects of Agricultural Research: The Case of Cotton Research in Sao Paulo, Brazil. *American Journal of Agricultural Economics* 54:557–569.

Consultative Group on International Agricultural Research. 1976. *International Research in Agriculture.* New York: United Nations Development Programme.

Duncan, R. C. 1972. Evaluating Returns to Research in Pasture Improvements. *Australian Journal of Agricultural Economics* 16:153–168.

Feistritzer, W. P., and Kelly, A. F., eds. 1978. *Improved Seed Production.* Rome: FAO.

Gilmour, J.S.L. 1969. *International Code of Nomenclature of Cultivated Plants.* Utrecht, Netherlands: Bohn, Scheltema, and Holkema.

Government of India. 1968. *Seed Review Team Report.* New Delhi: Department of Publications.

Griliches, Z. 1958. Research Costs and Social Returns: Hybrid Corn and Related Innovations. *Journal of Political Economy* 66:419–431.

Himes, J. R. 1972. The Utilization of Research for Development: Two Case Stud-
    ies in Rural Modernization and Agriculture in Peru. Ph.D. thesis, Princeton
    University.
Jugenheimer, R. W. 1976. *Corn: Improvement, Seed Production, and Uses.* New
    York: Wiley.
Moseman, A. H. 1970. *Building Agricultural Research Systems in the Developing
    Nations.* New York: Agricultural Development Council.

3

The Seed Program Starts:
Initial Seed Multiplications

Crop research is not a seed program. It is, however, the foundation on which a good seed program is built. The seed program should start as soon as, or perhaps before, decisions are made to introduce, recommend, or promote the use of a new variety. Getting the initial seed multiplications made and supplying the seed to others for further increase frequently is the weakest link in a seed program. Plant breeders ask, Is this really our responsibility? Research station managers ask, Should our land be used for this purpose? Administrators ask, Should we distribute seed directly to our farmers from the research station?

Private seed enterprises that have plant breeding programs usually develop ways to answer these questions and manage initial seed multiplications within their own organizations. However, answers to such questions frequently need to be found in developing countries that have a high proportion of publicly supported crop research.

THE FIRST CRITICAL STEPS

The development of new varieties has value only if seed becomes available. The vital first step toward producing adequate seed supplies of a variety is to decide how and where to make the initial multiplications of the first few kilograms of seed and who will be responsible for making the multiplications. Three important considerations in making this decision are (1) the need to maintain the characteristics of the variety as described, (2) the need for a mechanism of seed maintenance and multiplication, and (3) the value of cooperation.

Importance of Variety Maintenance

A variety name promises certain genetic qualities in the seed a farmer buys. When a plant breeder develops a variety he must ensure that the characteristics he has bred into the variety are genetically stable. It is, however, the further obligation of a seed program to maintain those genetic characteristics in the variety while multiplying its seed.

Most crop species reproduce sexually, so each time seed is produced there is a chance that some changes may occur. The environment can affect plants, or chance contamination with pollen from other varieties can cause change. Furthermore, it is rarely possible to identify a variety solely by looking at seed or even at a seedling. Indeed, a variety description must be made by observation of the growing plants from the seedling stage to maturity. For example, a wheat variety might be described through its reproductive cycle in terms of seed color, seedling plant habit, the time of spike emergence, straw length, and the characteristics of the mature spike.

When a farmer plants the seed of a variety, he assumes the essential characteristics of the variety have been retained throughout the preceding multiplication steps. A seed production program must be designed so that this assumption will be correct.

The Mechanism of Seed Maintenance and Multiplication

When a plant breeder releases a new variety for use, he must retain and maintain a small supply of genetically pure seed of the variety for further multiplications. The small number of plants used for the maintenance of the variety are usually called parental material. The plant breeder may make some intermediate multiplications, but eventually enough seed is available to pass on to others for further multiplication. Plant breeders should not be responsible for supplying seed directly to farmers.

Seed of the maintained variety is multiplied through a few cycles of reproduction—the initial seed multiplications. In seed certification programs, the final classes of the initial seed multiplications are identified in different ways, but the terms Breeder Seed and Basic Seed are used in this book. Basic Seed is the last step in the initial seed multiplications and is intended for the production of Certified Seed (Figure 2). The generation preceding Basic Seed is Breeder Seed (in the OECD seed schemes, the generation preceding Basic Seed is called Pre-basic Seed; see Appendix E).

Variety maintenance and each cycle of multiplication must be observed carefully by technical experts who remove any plants showing characteristics not typical of the variety. In well-run systems, the staff keeps careful records of the history of individual plants or seed lots and

Figure 2. Simplified multiplication sequence for self-pollinated crops

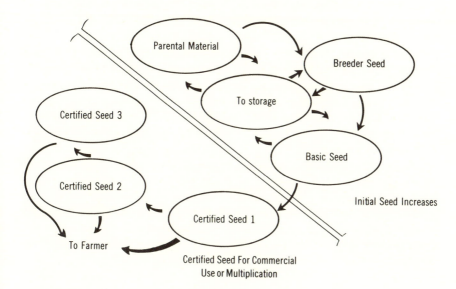

has an intimate knowledge of the characteristics of all varieties in the program.

Maintenance and initial seed multiplications of all varieties are necessary whether the varieties result from a public or a private plant-breeding program. Both types of programs face the same technical problems, but there are often special problems of organization in the public sector that must be dealt with if the results of a plant breeding program are to give the maximum benefit to agriculture. Most of the remainder of this chapter is intended mainly as a guide on how to handle initial seed increases in the public-sector programs.

International Cooperation

Neighboring countries can cooperate in varietal maintenance and initial seed increases so that each country need not maintain every variety it uses. Responsibility can be shared to ensure an adequate supply of Basic Seed for further multiplication within each country. The degree of cooperation obviously depends on political factors, but international agreements such as the seed schemes of the Organization for Economic Cooperation and Development (OECD) make such cooperation easier.

The international agricultural research institutes (see Chapter 2) do not normally release and maintain named varieties. Thus, they do not attempt to multiply enough seed for large-scale production. However, their close contacts with plant breeders in many countries make it possible for them to provide help and advice. A recent example was the multiplication of seed of the Korean rice variety Tongil (see Appendix B). The International Rice Research Institute helped with the initial seed multiplications of Tongil and also arranged contacts between Korean officials and Philippine officials to facilitate additional multiplications so the variety could be introduced to Korean farmers more rapidly.

Administrators need to be on the alert for ways to achieve high seed multiplication rates and thus reduce the time required to supply farmers with seed of new varieties.

ORGANIZING TO MAINTAIN AND MULTIPLY VARIETIES

The organization of variety maintenance and initial seed multiplication is shaped by the involvement of the plant breeder. Following maintenance, there generally are one or two further steps (intermediate stages) in the multiplication cycle before Basic Seed is produced under a seed certification system. Only a small amount of seed comes from the maintenance program, and it must be multiplied so that the required quantity of Basic Seed or its equivalent can be produced.

Broadly speaking, there are three ways the plant breeder may be involved: (1) the plant breeder is wholly responsible for the maintenance of varieties, for the intermediate stages before Basic Seed, and for producing Basic Seed; (2) the plant breeder works with a separate group that maintains the breeder's varieties, produces the intermediate stages, and produces Basic Seed; (3) the plant breeder provides advice and supervision to a totally separate enterprise that produces all intermediate stages and Basic Seed.

To decide which method of organization is most suitable, short-term needs must be clearly distinguished from long-term needs. All decisions should allow for sufficient flexibility so it is always possible to move from one organizational plan to another.

Plant Breeder Wholly Responsible

Obviously, making the plant breeder wholly responsible for maintenance and multiplication is the simplest decision. The plant breeder knows his variety better than anyone else, and it will take time for others to acquire this knowledge. In developing countries, plant breeding often is a very personal endeavor. One person makes the cross, selects potential varieties from the resulting plants, and follows them through varietal testing—perhaps even to trials in farmers' fields. In this situation the plant breeder usually is also responsible for maintenance of the variety and its initial multiplication. This type of organization often is adequate for seed programs in stage 1. The great disadvantage is that the more good varieties a plant breeder produces, the more time he must spend on maintenance and multiplication work, which, in turn, means less time to breed new varieties.

To ease a plant breeder's work load, an administrator can assign a specialist in variety maintenance and initial seed multiplication to the plant breeding department. However, if the "specialist" is chosen from people who have been working in the plant breeding program, he will have no special skills in seed multiplication or variety maintenance. Moreover such a person may continue to be obliged to devote a great deal of time to plant breeding tasks.

The plant breeding staff must appreciate the role of a seed maintenance and multiplication technologist. An individual with the special skills and the aptitude for such work may be difficult to find (Appendix A contains a sample job specification), and special training will often be necessary (see Chapter 7). Therefore, once such an individual has been found, those special skills should be fully employed.

A variation of the individual approach to plant breeding is to use teams. Breeders making crosses may advance them through two or three generations, but the subsequent progenies are evaluated by a multidisciplinary team. Still another group of technicians evaluates the variety in trials. This method generally offers a more objective assessment of the potential of a cross. The resulting product of the team approach will still require a maintenance and multiplication program. Although a plant breeder initially knows a variety better than anyone else and could continue to do maintenance and multiplication work, the team approach lends itself more readily to the organizational patterns in alternatives 2 and 3.

Plant Breeder Plus a Specialized Group

There may be numerous maintenance and multiplication programs in countries where several plant breeders are at work. Economies are possible by combining programs to provide maintenance and multiplication facilities for several plant breeders through a separate group. This group may be a department within a research station where the plant breeders are working or a separate organization. The need for such a group depends on the stage of development of both the plant breeding program and the seed program.

The creation of a separate department within a research station will give status to the work of variety maintenance and initial seed multiplication. Under this type of organization, the plant breeder's task is to identify new genotypes that can be developed into improved varieties, and members of the seed maintenance and multiplication group specialize in maintaining and multiplying released varieties. When the station has breeding programs for several crops, such a group should be responsible for variety maintenance and multiplication of all the crops, with guidance from the various plant breeders. The professional background and skills required are broadly the same for most crops.

If both the maintenance and multiplication group and the plant breeding program are departments of a research station, the plant breeders should have no difficulty providing advice and practical help for variety maintenance.

Plant Breeder and a Basic Seed Enterprise

By the time a seed program reaches stage 3 or 4, it is often desirable to create a separate Basic Seed enterprise for variety maintenance and initial seed multiplication in order to increase operational and financial flexibility. A separate enterprise can adjust quickly to changing needs for particular varieties. It can balance shortages of Basic Seed in one area of the country with surplus supplies in another. If climatic conditions hamper seed production of a locally bred variety, the enterprise could arrange to send small quantities of Breeder or Basic Seed abroad for multiplication in a more favorable climate. If large quantities of Certified Seed are being purchased abroad, it might be more economical to import and multiply the Breeder or Basic Seed instead.

As the Basic Seed enterprise becomes a major element in work on seed, recruiting and maintaining a specialized staff will become easier. Staff training must be recognized as fundamental to the enterprise's development. The staff should be trained to advise seed growers and seed enterprises on the production and processing of new varieties that are introduced into the program.

The responsibilities of a new Basic Seed enterprise must be carefully defined and a partnership with the plant breeders must be created or the work will fail. Especially when new varieties are released, plant breeders must help the staff of the new organization. With time, the technicians will learn to handle older varieties with little guidance from the plant breeders. Technicians from the Basic Seed enterprise can help the plant breeders with the multiplications prior to Basic Seed. A partnership in which both plant breeders and technicians feel mutually responsible will lead to efficient work.

It is wise to start on a modest scale and make provision for the enterprise to expand. For example, a new enterprise might start producing Basic Seed of the most widely grown crop. Its activities eventually could be broadened to include the earlier stages of multiplication or variety maintenance, or it could be made responsible for production of Basic Seed of other crops.

Basic Seed production and distribution also should be closely but informally linked with the next stages of multiplication. However, a Basic Seed enterprise should not attempt to assume responsibility for Certified Seed or commercial seed production (see Chapter 4).

METHODS AND FACILITIES

Since seed from initial seed increases is so vital to all subsequent multiplications, the following factors concerned with its maintenance and multiplication must be clearly planned: how each variety is to be maintained, the use to be made of long-term seed storage in the system, the number of multiplication cycles required, the amount of seed needed to support later multiplications, and the kind of facilities required for maintenance and multiplication. Development of such a plan requires the combined efforts of administrators, leaders of crop research programs, key personnel involved in the work, and people from the seed enterprises responsible for subsequent multiplications of seed.

Methods of Variety Maintenance

Maintenance of a variety can be accomplished by annually growing a small number of plants that represent the variety and harvesting seed from selected plants. Or, when a new variety is released, a quantity of seed can be produced and placed in long-term storage. A small portion of this seed is then withdrawn each year to start a new multiplication cycle. These two systems may be combined. The plant selection process can be repeated every three to five years, and the bulk seed can be stored for use in the intervening years.

In systems that use plant selection the genetic requirements of a variety may differ, but it is important to distinguish between variety maintenance and variety improvement. The former seeks merely to eliminate off-types and maintain the variety as tested; the latter is a plant breeding activity that ultimately should lead to the establishment of a new variety—for example, introducing new disease resistance into an existing variety.

The seed maintenance technologist must maintain progenies in such a way that the origins of all intermediate multiplications and Basic Seed can be traced. Maintenance involves some system of individual plant selection followed by a "trueness-to-type" test of the selected plants. These tests vary, but basically they depend upon evaluating seed samples, if necessary in field plots, before the next step is taken in the the seed supply chain. The OECD guidelines can be useful. Seed from selections deemed true-to-type is bulked and used to propagate the variety.

When plant breeding procedures change, maintenance methods may have to change too. For example, multiline varieties require new methods and increased attention by plant breeders as they maintain and blend the lines. Similarly, composite varieties of crops such as maize require highly diligent maintenance and care in the initial seed muliplication. The variation inherent in composites makes identifying admixtures more difficult than in more uniform varieties. The maintenance and multiplication systems must be adaptable.

Use of Storage

To maintain varietal purity, plants should be grown and selected as infrequently as possible during the "life" of a variety because of the chance of accident each time plant selections are made. The initial bulked material identified as true-to-type should be preserved for as long as possible. If facilities for long-term storage are available, the bulked seed can be stored so that subsequent multiplications can come from the initial multiplication for perhaps five to ten years. At the end of that period, the variety might be obsolete. If long-term storage is not possible, the alternative is to multiply only as often as necessary to maintain seed viability.

Proper facilities are essential for long-term storage of seed. Failure in storage can be disastrous if the seed of experimental or improved varieties—the result of expensive plant breeding—is lost. Providing the plant breeder with help for managing seed storage may be a more efficient use of funds than employing additional field-workers for variety maintenance. The storage facility may also be useful to other individuals in the breeding program. Whether or not a storage facility is built will depend upon the resources and skills a country has, the number of varieties in the program, and the amount of seed needed.

Multiplication Cycles

How many multiplication cycles will be needed to produce Basic Seed depends upon the genetic background of the variety and the importance of the variety to agricultural production. For example, because of the genetic background of F_1 or other hybrids, new cycles must be started each year to provide the quantities of Basic Seed required annually. But for many self-fertilized species, it may not be necessary to start the cycle every year because some of the later generations can be remultiplied easily. Figure 3 illustrates two possible systems.

Determining Requirements

The total requirements for the varietal maintenance and initial multiplication programs are related to the number of crops and varieties to be handled. For each variety and each stage of multiplication, estimates must be made of the quantity of seed needed, the frequency with which it must be produced, and the amount to be held in reserve. These estimates will depend upon the species, the expected popularity of the variety, the multiplication factor that may be expected from one generation to the next, and the number of multiplications that follow production of Basic Seed. The estimates should allow for production that may be rejected because of impurities and for losses due to natural causes such as crop failure or poor harvest conditions. The FAO manual *Improved Seed Production* gives several methods of estimating expected needs. In all stages, seed production areas or seed lots should be ruthlessly rejected if they do not meet quality standards. The economic consequences of rejection at early stages are far less than later rejection when larger areas and many farmers may be involved.

For this reason, production targets should not be rigid. Sometimes plots are not rogued because the yield might be reduced too much. Although the plot supervisor may achieve the target yield and thus avoid reprimand or monetary penalties, the poor seed quality is likely to cause trouble in later multiplications.

Meeting Basic Requirements

The maintenance of a variety and subsequent multiplications to Basic Seed require suitable land and equipment, a seed processing plant capable of handling small quantities of seed of individual varieties, storage facilities, systematic work, adequate records, and labeling.

Land and Equipment

Since plant breeders must be involved, variety maintenance, and probably initial multiplications, should be done on or near the research sta-

72

Figure 3. Systems of maintaining and multiplying varieties. In either system, any stage, except for variety maintenance in season 1 and the multiplication to Basic Seed from it, can be eliminated if sufficient long-term storage is available. Providing a seed storage facility can be regarded as a partial substitute for field work. In the intermittent system the intervals can be lengthened or shortened depending on species and variety. The broken lines from Breeder Seed to variety maintenance represent the possibility of making selections of typical plants in the Breeder Seed production plot for use for variety maintenance in the following year.

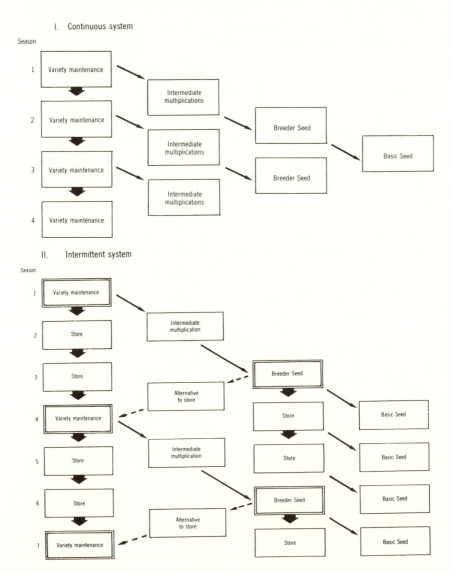

tion where the plant breeder is working. Enough land area must be available to allow for a rotation so there are suitable intervals between crops. An efficient use of limited seed supplies in these early multiplications requires low planting rates to achieve the greatest multiplication rate per seed sown rather than maximum production per unit area. The research station can justify production on this basis.

One difficulty common to research stations is that the volume of work is usually so great that sowing time may be delayed, crop culture may be poor, and yields may be low. Adding variety maintenance and initial seed multiplication work to a station's tasks should not be allowed to burden the farm staff or equipment. But neither should these added responsibilities cause other research-station work to be neglected.

When Basic Seed is produced, larger areas and greater quantities of seed are involved, and the research station may not have enough land. To ensure proper quality control, the *first* multiplication of Basic Seed of new varieties should be done at the research station. A few farmers can be selected to multiply established varieties on contract. Having Basic Seed grown on farms in several locations is advantageous because it spreads risk. A sound contracting system must be developed, however, and seed production areas must be closely supervised.

Some special field equipment will be needed (see Chapter 8). For initial multiplications, planting can be done with any equipment that will maintain purity. For instance, instead of large seed drills, simple equipment should be used because it is easier to clean between seed lots. Equipment for harvesting, threshing, cleaning, drying, and processing must be chosen with care because contamination is more likely during these steps than elsewhere in the multiplication cycle. Small units that can be easily cleaned are desirable. Use of equipment that is too complex for the staff to handle properly is certain to result in purity problems.

Drying and Processing

Because seed must be adequately dried, processed, packaged, and stored, these activities should be concentrated in at most a few locations around which seed growing can be organized. Distribution of Basic Seed will require adequate transport.

The size of the seed processing and storage plant should match the size of the general program—the number of crop species, number of varieties, and quantities of Basic Seed required, which may range from a few kilograms to several tons. Two types of processing equipment will be needed: one type to handle and package small quantities of seed in the early stages of the multiplication work and large equipment for the Basic Seed stage. In selecting equipment, efficiency in terms of output per hour is far less critical than how easily the equipment can be cleaned between seed lots.

Storage

The processing plant must have adequate short-term seed storage facilities and, possibly, intermediate- and long-term storage facilities (for use in the intermittent system illustrated in Figure 3). All stored seed requires proper attention and careful handling. Storage containers must be easy to clean thoroughly before use. If seed is stored in bags, only new bags should be used (details on seed storage are given in Chapter 8).

Work, Records, and Labeling

The opportunity for mistakes during seed processing and storage are great, and the consequences are far-reaching and costly. The staff must have special training that emphasizes that methodical work is more important than speed. And the inventories and records must be well kept so that the records, together with the labeling system, allow the complete history of every seed lot to be traced.

POLICIES FOR THE USE OF BASIC SEED

In planning the production of Basic Seed, it is necessary to consider how to supply farmers with authentic seed of varieties. How much and how often Basic Seed, or its equivalent, should be produced will depend upon the policy for its use.

One possibility is to supply Basic Seed, or its equivalent, in small quantities to a large number of farmers who then multiply the seed without supervision and supply it to their neighbors in the following season. The farmer-to-farmer movement of seed results in widespread use of a variety with little official assistance. If, however, farmers produce poor seed because they are careless or do not have proper facilities, some risk exists that weeds and diseases will be spread with the seed.

A second possibility is to supply Basic Seed, or its equivalent, to selected seed growers and seed enterprises who then multiply it and market (or arrange for the marketing of) Certified Seed or commercial seed to farmers. This system can stimulate interest in seed production and marketing among many individuals and enterprises. An equitable policy for allocating Basic Seed of new varieties must be devised which will ensure that those with experience in seed multiplication can participate in the program on a continuing basis (Appendix A contains suggestions for an allocation system).

A third possibility is to supply Basic Seed, or its equivalent, only to

The growth of initial seed increase components

INITIAL SEED INCREASE	Stage 1	Stage 2	Stage 3	Stage 4
Production by plant breeder	**	**	**	**
Production by plant breeder plus specialized group		***	****	
Production by plant breeder plus separate basic seed enterprise			**	********
Production by seed enterprise with own research and seed increase			**	****
Seed supplied to farmers	*	**	*	
Seed supplied to seed growers		**	**	**
Seed supplied to seed enterprises		**	***	******

public farms and units, and they, then, have the responsibility to multiply and distribute it to farmers. This system does not encourage seed growers or seed enterprises to produce and market seed.

Considerations for the Administrator

Administrators should distinguish between short- and long-term needs. The immediate problems in supplying new varieties to farmers may be overcome by arranging for the production of Basic Seed, or its equivalent, and leaving the rest to hundreds of farmers. This system may be the least complex and the least costly in areas where no other seed multiplication system exists. However, there is some risk that all of the seed will not be planted and that the seed multiplied and supplied to neighboring farmers will not be of a good quality. In the long run, more comprehensive arrangements may be needed to facilitate a rapid changeover when new varieties are produced and to ensure that the established varieties are properly maintained until they are replaced. In addition, cross-fertilized species of forages, many vegetables, and hybrids do not lend themselves to widespread seed multiplication by many farmers. For such crops it will be necessary to use one of the other systems of multiplication.

Supplying seed to farmers to multiply for their neighbors can lead to a system of regular seed growers and to seed enterprises and marketing groups. Chapters 4 and 6 deal with various possibilities for organizing seed enterprises and marketing groups. Although it takes time and special effort to get such groups functioning, they can add strength, stability, and momentum to a seed program.

Sometimes government multiplication—the third option—is the only possibility. Nevertheless production of Basic Seed (or its equivalent), subsequent multiplications, and effective marketing remain as important as in the second option. The amount of Basic Seed needed, however, can probably be estimated more easily in this situation. On the other hand the government will have to take more responsibility for the overall program and bear a greater financial burden.

Pricing Basic Seed

Because Basic Seed, or its equivalent, is limited in quantity, it should be carefully multiplied to achieve the maximum increase possible. This factor needs special consideration when the seed is being priced and allocated. People take better care of something if it has been a bit costly. Thus, to charge more for Basic Seed, or its equivalent, than for commercial seed or Certified Seed is a common practice in many programs.

Initial Seed Multiplication: Major Policy Points

1. Amount of seed of a new variety to be multiplied by public research institutions before the seed is distributed to others for further multiplication and use
2. Organizational pattern to be used by crop research programs to maintain and multiply varieties
3. Whether a Basic Seed enterprise is to be separate from the crop research program, its role, and the limits of its responsibility
4. Whether the Basic Seed enterprise is to be partially or totally self-sufficient
5. Use and allocation of Basic Seed (or its equivalent) to farmers, seed growers, seed enterprises, or public-sector farms with respect to different crops

---

Usually Basic Seed is more expensive to produce than other seed because smaller volumes are involved, extra roguing is needed, there are added storage requirements, and the risks of production are higher—especially with new varieties. Since Basic Seed production often can be self-sufficient by the time a program reaches stage 4, and perhaps provide revenues for research, it is important that all costs of production be identified and recognized when pricing Basic Seed (more details on pricing are included in Chapter 6).

Basic Seed and Seed Certification

Occasionally seed from the initial increases is not pure enough to be called Basic Seed; therefore, "its equivalent" may be all that is available. It may be necessary to use this seed, but it would not enter a seed certification system.

When seed is pure enough to satisfy the seed certification requirements, it should be officially called Basic Seed and handled through the seed certification system. Basic Seed production should be inspected by the seed certifying authority, but, if manpower is scarce, the seed certification authority may have to rely heavily upon the plant breeder and the Basic Seed group to help inspect production plots and fields. As the seed certification staff grows and gains in experience, it normally assumes responsibility for "official" inspections of all Basic and Certified Seed production. A good working relationship between those responsible for Basic Seed production and those in charge of the certification scheme is necessary for success (seed certification and its role are discussed further in Chapter 5).

REFERENCES

International Rice Research Institute. 1976. How Tongil Triggered a Korean Rice Revolution. *IRRI Reporter* no. 3.

Lewis, C. F. 1970. Concepts of Varietal Maintenance in Cotton. *Cotton Growing Review* 47:272–284.

Organization for Economic Cooperation and Development. 1969. *Guide to Methods Used in Plot Tests and to Methods of Field Inspection of Cereal Seed Crops.* Paris.

_____. 1973. *Guide to the Methods Used in Plot Tests and to the Methods of Field Inspection of Herbage Seed Crops.* Paris.

_____. 1977. *OECD Scheme for the Varietal Certification of Cereal Seed Moving in International Trade.* Paris.

_____. 1977. *OECD Scheme for the Varietal Certification of Herbage and Oil*

Seed Moving in International Trade. Paris.

_____. 1977. *OECD Scheme for the Varietal Certification of Maize Seed Moving in International Trade.* Paris.

_____. 1977. *OECD Scheme for the Varietal Certification of Subterranean Clover and Similar Species Moving in International Trade.* Paris.

_____. 1977. *OECD Scheme for the Varietal Certification of Sugar Beet and Fodder Beet Seed Moving in International Trade.* Paris.

_____. 1977. *OECD Scheme for the Varietal Certification of Vegetable Seed Moving in International Trade.* Paris.

4
Building the Seed Supply

Producing enough seed is one of the most difficult problems facing administrators. Should seeds be produced locally or imported? Who should produce the seeds—the farmers, the government, or seed enterprises? How can a seed enterprise be started and what is involved in its management? What is actually involved in getting seed produced? Can foreign seed firms help get a commercial seed industry started? As answers to these questions are found and clear policies are established, a country's seed program takes a giant step toward success.

IMPORTATION

It may take a new seed industry several years to develop the production capacity to meet national needs. In the meantime, importing seed can build supplies quickly if varieties suitable for the country are available abroad. Importing seed may also be wise for minor crops or for crops whose seed is difficult to produce locally. Vegetables, forages, and sugarbeets, for example, are harvested before they set seed so the areas in which they are grown commercially may be poor areas for seed production. Even countries that have a well-established seed industry import certain categories of seed. The United States, for example, imported over three million kilograms of agricultural seed from approximately twenty countries in 1976.

There are, however, dangers in relying solely on imports if the seed can be produced locally. Imported supplies may be disrupted by political changes; unfavorable growing conditions abroad may result in shortages; outbreaks of insects and diseases abroad may force the importation of seed to be banned. Furthermore, complete reliance on imported seed may lead to a dependence on foreign technology. The growth of a local commercial seed industry could be stunted by the availability of inexpensive imported seed; thus, seed of varieties developed through indigenous research might never be produced.

In attempting to protect the local seed industry or to preserve foreign exchange, some countries restrict or prohibit the importation of seed. These policies overlook the immediate benefits agricultural production can get from importing seed of improved varieties. Moreover, a lack of imported seed may inhibit the production of vegetables and other special crops. Since such policies seldom differentiate among types of seed, the development of a country's seed enterprises may be hampered by the inability to import seed for further multiplication. And a restrictive policy often limits the number of importers and suppliers, increasing the farmers' vulnerability to price fluctuations or disruptions of supplies.

Some countries enact special legislation to control seed imports because of a concern that poor quality seed, seed of unadapted varieties, or seed carrying pests or diseases will be dumped on the local market. If imported seed meets the standards of the domestic market, special legislation should not be necessary.

Policies that strengthen the local seed industry will lessen a country's dependence on imported seed and will be more productive than policies that restrict imports. This is especially true for many cereal seeds as, for them, freight costs place imported seed at a disadvantage.

LOCAL PRODUCTION

In most countries where seed programs are in stages 1 or 2, only small amounts of seed are produced by research stations, government farms, and individual seed growers. Increasing the seed production capacity from this fairly low level is difficult. It requires persistent effort and an ability to generate interest in the seed program among agricultural leaders. The focal point should be the establishment and growth of seed enterprises in the private or public sectors, or both.

Before considering how to stimulate local production of seed, the current seed production capacity and resources of the national industry should be evaluated. A realistic appraisal of the farmers' demand for improved varieties is also needed. It is easy to hypothesize a demand when research trials give favorable results, but the actual demand for seed may be quite different (Chapter 6 deals in more detail with assessing seed demand).

Production and Processing Capacities

If the demand for seed exceeds the supply, the existing production and processing capacities must be reviewed before additional facilities are planned. To expand seed production, the first priority is to achieve maximum seed yields. Increasing the yield of seed per hectare will increase supplies and lower costs. Producing seed in new areas with more favorable growing conditions can result in more reliable and higher yields. A move to new areas may also be warranted to obtain a better supply of labor at peak periods or to be able to make use of irrigation.

In processing plants, phases of the operation that restrict output should be scrutinized. It may be more economical to purchase new equipment, remodel a plant, or add more storage capacity to an existing plant than to construct another processing plant at a new location. If a need does exist for more seed processing plants, risks can be spread by locating plants in different climatic zones, even though several small plants may cost more than a few large ones. The size of the seed production area and the capacity of the processing plant should be planned together.

The costs of expanding the total seed processing capacity to meet the demand for seed cannot be estimated until the additional needs are clearly identified. So many factors are involved that the capital cost of a processing facility may vary from a hundred thousand dollars for a small, modestly constructed plant that handles seed that requires no special processing, to millions of dollars for a large, centrally located plant that handles several kinds of seed.

Money and Manpower

Compared with other industries, the seed industry is not capital intensive. The investment required for production facilities can be easily justified because the benefits realized by the farmer, and the economy, from the use of seed of improved varieties far exceed the cost. This is true whether the investment is made in the public or private sector. Few other investments benefit the agricultural economy as much.

On the other hand, the need for working capital is great, especially around harvesttime when the seed enterprise purchases the growers' seed crop. Although the price paid for unprocessed seed varies according to the crop, premiums paid for the seed of self-pollinated crops generally are 10 to 20 percent above the grain price. For hybrid seed the premiums are much higher. Since unprocessed seed purchased at harvesttime might not be sold until the following year, a small seed enterprise, handling five thousand tons of hybrid seed, might require credit for nine to ten months of the year with a peak debt well over a million dollars soon after harvest.

A basic policy decision in establishment, management, and operation of a seed enterprise centers on its opportunity to make a profit. If the op-

Letterheads of some seed enterprises in developing countries

portunity does not exist, private seed enterprises will not be started or survive. Joint or public seed enterprises can be subsidized and can operate without a profit, but they may be a burden on the national treasury. Many countries have found that allowing seed enterprises to earn a reasonable profit is the easiest way to stimulate seed production and marketing. Profit is one way an enterprise can accumulate capital for further investment and growth (Chapter 6 discusses break-even and profit factors in seed pricing).

Enlarging the national seed production capacity will depend in part on the availability of trained personnel. Too often, the individuals selected have skill in only one phase of a seed enterprise. The seed industry requires a high level of managerial, financial, marketing, and technical skills variously combined in key positions (personnel development and staffing are discussed in Chapter 7).

To summarize, the characteristics of seed enterprises in many countries present a series of contrasts. Although the capital requirements are not particularly high, working capital needs can be burdensome. The enterprise does not have to own much land, but competent seed growers are essential. Except for the breeding programs, the industry is not labor intensive, but an adequate pool of temporary workers must be available for peak periods. It is an industry of small enterprises, each one requiring high levels of managerial and technical skills from its key employees. Many of these skills take years to acquire.

TYPES OF SEED ENTERPRISES

Some seed enterprises are wholly private (family companies, for example), some are wholly public (national seed companies), and some seed enterprises have various mixtures of public and private participation. Five approaches for developing seed production capacity are discussed below (and see Figure 4). Some are more likely to be acceptable than others for reasons such as a nation's ideology, economic factors, and the structure of the existing industry; often, more than one type of seed enterprise exists within a country.

Private Enterprises

In many countries the commercial seed industry has developed through private seed companies. Such seed enterprises may take various forms: individual holdings, partnerships, corporations, cooperatives, and associations structured within the commercial laws of the country. Private enterprises are primarily responsible to their shareholders or

Figure 4. Alternative methods for developing seed enterprises

Alternative	Crop Breeding	Initial Seed Increases	Commercial and Certified Seed Production	Internal Quality Control	Marketing
Private Only	Private	Private	Private	Private	Private
Private with Normal Government Assistance	Public	Public	Private	Private	Private
Private with Maximum Government Assistance	Public	Public	Private	Private	Private
Private with Maximum Government Assistance Plus. Joint Ownership	Public	Public	Joint Ownership	Joint Ownership	Joint Ownership
Predominately Public Ownership	Public	Public	Private / Public	Public	Public

Public [ ]   Private (○○○○)   Joint Ownership (◉◉◉)

members, though their activities are conducted in accord with seed and other trade legislation. These enterprises are often independent of direct government support and conduct research to develop new proprietary varieties. They multiply, process, and distribute seed of their varieties and sell it to farmers.

The government is mainly concerned with protecting the agricultural industry and the farmer-consumer through legislation. Such legislation is enforced to prevent the introduction of pests and diseases, which could cause economic damage, and to protect the consumer against the purchase of inferior seed (Chapter 5 discusses this subject further).

Among the advantages of a competitive private seed industry are its flexibility to meet changing demands, its cost efficiency, the role it can play in helping the farmer buy better seed and produce high-yielding commercial crops, and its ability to provide various services to the farmer.

Private seed organizations have some disadvantages too. They mainly specialize in those crops that will earn sufficient profits to justify a research and development program. Therefore, even when a private commercial seed industry has developed, government-supported research on many crops is still needed. A more important drawback is that a private commercial seed industry evolves gradually and too much time may elapse before a nation's need for seed can be satisfied unless the government stimulates the industry's development.

Private Enterprises with Partial Government Assistance

Most governments help the commercial seed industry through research. Public research institutions are often the major source of new varieties, especially for crops not covered by private research. These public institutions can give vital support to a commercial seed industry by releasing germplasm that may provide a base for research in the private sector. Multiplication of Breeder and Basic Seed by public agencies is a common method of indirect support to the private sector in several countries. Normally, this seed is sold to the private sector for further multiplication and sale.

Governments of developing countries can encourage the development of private seed enterprises in other ways. Even if the climate for investment in the private sector is favorable, potential seed industry investors may need incentives such as credit at low interest rates, grants for capital investment, and special credit for financing seed inventories.

Although the capital requirements for a seed enterprise may be modest in comparison with other businesses, they may be too large for the most progressive, interested people to become involved. Moreover, the seed business is exceptionally risky. A seed crop is more difficult to produce than a commercial crop, and it suffers correspondingly greater risks from the vagaries of weather. Recognizing these needs, some governments have developed special lines of credit for their seed industries.

The seed industry does not usually require subsidies, though temporary subsidies are sometimes used to encourage the production of seed of certain crops until the volume of seed sold is large enough to earn sufficient profits to offset the original investment. This situation might arise when the production of seed of a nationally important crop does not offer an attractive investment opportunity to private seed enterprises. Such a subsidy policy may be more efficient for a government than becoming directly involved in seed production and marketing.

As in most businesses, pricing is critically important. Legislation and administrative orders must clearly distinguish between seed and grain so that taxes and other levies imposed on grain are not automatically applied to seed. Caution is needed to ensure that policies do not curtail the development of seed enterprises. Any form of price control will be detrimental to the production of seed in the private sector—whether direct price controls or indirect through subsidized, competitive government production.

The provisions of seed laws and regulations may stimulate growth in the private sector, or they may act as a curb. Plant variety protection legislation can encourage research in the private sector, but it can also have a restrictive effect by reducing the flow of germplasm exchange. If quarantine regulations and seed laws are too restrictive, they may inhibit development.

Private Enterprises with Maximum Government Assistance

Private seed enterprises can receive a wide range of government support that is still short of actual government investment but greater than the more conventional assistance described above.

A lack of facilities and equipment may be a major reason why individuals and groups hesitate to begin a seed production program. As a first step a government might modify foreign exchange restrictions to facilitate the purchase of field, seed processing, and storage equipment. A government might also provide for the import, installation, lease, or lease-purchase of equipment needed in seed processing and storage facilities. Because governments may not want to provide equipment as a grant, a lease-purchase agreement may be more practical.

If facilities already exist, leasing them may be the best solution when a government does not wish to sell and a prospective seed enterprise is reluctant to invest. Some governments have constructed warehouses and then leased them to seed enterprises.

A government seed development team can be helpful to those considering the seed business. A team might be composed of three persons: a seed production agronomist, a seed production and processing engineer, and a specialist in management, finance, and marketing. The team's activities could include (1) finding and assessing opportunities in seed production and marketing; (2) advising how to form a seed enterprise; (3) guiding production and marketing plans and providing technical assistance; (4) assessing seed processing and storage needs and assisting with the purchase, installation, and operation of equipment; (5) assisting seed associations; and (6) organizing educational meetings, field days, and special tours for seed growers and dealers. Educational tours and training opportunities abroad, for the team as well as for seed dealers in the private sector, could be part of the program.

While groups and individuals interested in seed production and marketing may have adequate business sense and financial resources, they may lack knowledge of seed production and technology. Government employees who have these skills or who have received special training could be temporarily assigned to seed enterprises to help initiate new programs.

A government-assisted seed development program can help seed suppliers by such measures as distributing lists of producers and suppliers of seed, seeing that seed producers and suppliers have clear information about the varieties being promoted, and planning seed production and marketing needs with private groups.

Other methods are needed when it is not feasible to form seed enterprises, areas are difficult to reach, or improved varieties exist for an area that lacks a mechanism for distributing seed locally. In these situations, the seed development team should find key farmers who can act as seed multipliers. In countries where the average farm is too small for efficient seed production, production should be encouraged among groups of progressive farmers whose land holdings are contiguous. These farmers can help test improved varieties, and, subsequently, they can multiply seed of the best ones for their neighbors. These seed multipliers should be informed of the best cultural practices for the varieties they are multiplying, the requirements for seed multiplication, the ways to maintain seed quality, and the importance of testing seed before distribution. Arrangements could be made to provide the seed multipliers with minimal supplies such as small cleaners, storage drums or facilities, fumigants, bags, and labels. Although this approach does not constitute a seed program for a country, it can meet short-term needs in some areas.

The objective of a maximum assistance program by a government is to stimulate seed production and marketing through as many individuals and groups as possible, but in line with actual seed needs. If an economic climate favors such a development and a research program is providing better varieties, seed production and marketing programs will become self-sustaining and adjust to future and expanding needs with little direct government involvement and ownership.

To sum up, the advantages of partial or maximum government assistance (compared with direct government participation discussed later) include

The responsibilities and risks of the seed production and marketing programs are spread among many individuals and groups who are closely tied to agriculture and who understand the farmers' problems.

The government does not become directly involved in seed produc-

tion and marketing activities from which it will be difficult to
disengage.

The resources of government are used to stimulate others to invest
their own talents and funds.

The future continuity of a program is ensured by the commitment
of leading farmers and individuals to seed production and
marketing.

Training will have a more lasting value since trained people will
not be transferred to other posts.

The existence of several seed enterprises will reduce production
risks, concentrate on local needs, minimize delays in getting seed
supplies to farmers, and reduce transportation costs.

The competitive forces developed will help ensure more efficient
performance and better seed quality over the long run.

The disadvantages of partial or maximum government assistance (compared with more direct government involvement) are

More educational effort is needed to involve and train many farmers
and other groups in seed production and marketing activities.

Providing good management for several seed production and marketing units will be more difficult.

Seed legislation must be developed to assure consumers of a seed
quality above specified minimum levels.

The initial investment in facilities and equipment may be larger
because more relatively small units will be formed.

Production and marketing units may remain too small and too scat-
tered.

National coordination of the seed production and supply programs
is more difficult.

Joint Public-Private Enterprises

When the private sector fails to achieve the needed seed production
and marketing objectives even with government assistance, more direct
government participation may be useful. Establishing joint seed enterprises involving both private- and public-sector capital could help
stimulate private involvement in a seed industry.

A joint seed enterprise can take various forms. It may include private
capital from individuals and business groups joined with capital from the
public sector. Such a joint enterprise might arrange for seed to be grown
on government farms or on private farms under contract to the joint seed

enterprise. Another kind of joint enterprise might combine farmer–seed growers and government farms (including farms controlled by agricultural universities and research institutions) with business and public capital. In such a structure the individuals and institutions that grow seed are joint owners of the seed enterprise. The objective is to bring private as well as public farming institutions into a harmonious seed production and marketing enterprise. Such a joint enterprise could be incorporated into a legal entity as a cooperative or company. More than one enterprise could, and probably should, operate in a country, depending on the kind and quantity of seed required and the availability of suitable areas where such units could be located.

Joint enterprises are advantageous in several ways:

> The combined resources of the private and public sectors are directly available to the seed industry.
>
> Government assistance, including technical guidance from research stations and agricultural universities, is fully available to the enterprise.
>
> The interests of the government, seed growers, industry groups, and others can be merged into an operating unit to achieve common objectives.
>
> The production of all crop seed can be accurately planned since the government and the private sector are jointly managing the seed enterprise (the private sector tends to concentrate on seed that has a higher profit).
>
> Trained and specialized farmers will continue to participate in seed production programs if both private seed growers and public-sector farms own the seed enterprises.
>
> The government is a business partner but does not have full control of seed production and marketing: the enterprise benefits from knowledge of government policy and information, and the government is able to carry on regulatory and quality control functions more objectively than if it were totally responsible for production activities.

Potential disadvantages of joint seed enterprises are

They may be difficult to form and manage.

Government involvement limits flexibility of operation.

Government nominees to the governing body of a seed enterprise, and the nominees' attitudes, may change frequently—especially if the government changes—creating problems of continuity and consistent management policies.

Government involvement can result in the adoption of politically motivated seed production policies that may not be economical.

The reliance of a seed enterprise on government may delay its becoming self-reliant and economically sound.

Direct government participation in a joint seed enterprise can create disparities that discourage development of seed enterprises that do not have government participation.

Government Enterprises

In some countries, the development of a private seed industry may be politically unacceptable or the private sector may fail to meet the full demand for all kinds of seed even with government assistance. Consequently, a government may establish a government-operated enterprise that contracts with private growers for seed production and markets seed through private dealers. Or there may be government seed enterprises that produce seed only on government farms and market it through government channels.

If a country's seed requirements are limited, the entire production program could be under one agency. However, in most countries the seed production and marketing requirements are extensive and complex, and more than one seed organization may be necessary.

Governments differ greatly in their ability to operate seed enterprises and marketing programs. The potential advantages of full government participation in seed enterprises often are not achieved. It should be possible to set up a government enterprise quickly to meet at least part of the seed demand. It should also be possible to plan a balanced production of seed of various kinds and ensure widespread distribution to all

The growth of seed supply components

BUILDING THE SEED SUPPLY	Stage 1	Stage 2	Stage 3	Stage 4
Seed importation	*	* *	* *	* * *
Production by research station, public sector farms and other seed growers	*	* * *	* *	* *
Production by seed enterprises with seed grower contracts		* *	* * * *	* * * * * * * *

users. If seed production and processing under government control were as efficient as under private management, and costs were comparable, the retail price of seed could be lower since no return on government capital is necessary. Furthermore, availability of working capital should not be a problem.

On the other hand, government seed enterprises may cause a seed program to suffer from the disadvantages of a monopoly system. A wrong decision by a few individuals—production of the wrong variety, use of the wrong cultivation technique, or similar errors—could affect all seed production in a country and jeopardize the seed supply. Such mistakes could seriously damage agricultural production. A second disadvantage is that government agencies often are unable to delegate decision-making authority to lower levels. For successful seed production, managerial decisions must be made promptly and on the spot. Third, rigid, departmentalized government machinery can prevent contact between various units or departments. Since seed production is subject to frequent and abrupt changes in varieties, weather, diseases, and pests, meaningful discussion between departments is necessary to avoid poor seed production. Fourth, a seed marketing program under government control is usually weak. As a result the government often suffers large losses or inferior and old seed is sold to farmers. Fifth, funds for expenditures often are not available in the amounts needed and at the right time.

Using the Alternatives

To summarize, there are five alternatives for expanding seed supplies.

1. Private seed enterprises that have their own research programs and total control of all seed multiplication and marketing functions
2. Private seed enterprises that receive partial assistance from government—such as publicly bred varieties, seed stocks for further multiplication, special credit concessions, and subsidies—without government interference in pricing
3. Private seed enterprises and seed production activities that get maximum government assistance such as equipment and building leases and lease-purchase arrangements, special uses of government staffs, marketing help, and measures to stimulate "seed multipliers"
4. Joint seed enterprises that involve both private and public capital
5. Government seed enterprises and seed activities that have only

government participation in all or parts of the seed production
and marketing programs

An administrator does not have to choose one type of seed enterprise
to the exclusion of all others. Many countries have wholly private seed
enterprises as well as ones partially assisted by the government. Mexico
has wholly private seed enterprises, ones receiving partial government
assistance, and government enterprises. Brazil has all types except joint
seed enterprises, although it is moving to eliminate government enterprises. India has had all five types; the Tarai Development Corporation
described in Appendix B was the first joint seed enterprise formed in India. Tunisia and Algeria have private seed enterprises receiving partial
government assistance as well as public enterprises; vegetable seeds, for
example, are handled primarily by private enterprises with partial
government assistance while much cereal seed is in the hands of government agencies. The Philippines has all types except joint enterprises; the
type involved varies from crop to crop. Kenya has all types of enterprise
except wholly private and wholly governmental. The Kenya Seed Company started as a private seed enterprise but now is a joint enterprise—the Agricultural Development Corporation owns shares along
with the original private shareholders (see Appendix B). Countries in
Eastern Europe, the USSR, and a few developing countries have only
government enterprises. Government enterprises in some countries produce Breeder Seed and Basic Seed, initiate commercial seed production
for special crops or for selected areas where other alternatives have not
been possible, or train personnel in seed growing, processing, and quality control for employment in other phases in the program.

Most administrators try to use resources in the most advantageous
combination to maximize seed production. The public sector frequently
becomes involved in crop breeding research, quality control regulation,
education, information, and planning. The most successful programs
have left the seed production, processing, and marketing to enterprises
involving the private sector. When a country has several types of enterprises, care must be taken to give all segments of the commercial seed industry equal opportunities to develop.

SEED ENTERPRISE—FORMATION, ORGANIZATION, AND MANAGEMENT

Recognizing that several alternative methods exist for organizing seed
enterprises, it is also important to consider what is involved in forming a
seed enterprise, the various organizational patterns possible, and the
management of an enterprise.

Forming a Seed Enterprise

A favorable economic climate is essential when forming a seed enterprise involving the private sector. Governments can enhance this process; otherwise, the formation of private seed enterprises occurs slowly.

Establishing a nation's first seed enterprise requires considerable effort by its management and government personnel. Help from a government seed development team can be extremely valuable. If the first enterprise succeeds in producing and marketing high quality seed, demonstrates profitability, and establishes a good reputation, others will be encouraged to follow.

The simplest seed enterprise to form is one limited to a family operation or a partnership. Because they are small, few people are involved and decision making is less complex, but many small enterprises may be needed to achieve national production objectives.

The formation of a cooperative or a company with many members or shareholders is more difficult but worthwhile because of the potential for a larger production capacity and the advantages of a concentration of resources and management. In forming such groups, areas of common interest among the participants should be identified; the immediate and long-term economic prospects for seed production and marketing in the region should be determined; the potential contribution of each person or group to the seed enterprise should be assessed; the method of incorporation that best suits the group should be decided upon; the physical and financial resources required in the short- and the long-term should be identified; staffing needs and availability of management personnel and technical leadership should be considered; the kinds and amount of external assistance needed, if any, should be determined; and a systematic plan for beginning the enterprise should be developed.

Government seed enterprises that are autonomous (although all shares are held by the government) have more flexibility and greater advantages than operations organized within government departments. The considerations involved in setting up an autonomous, government-controlled seed enterprise would be similar to those outlined for setting up a cooperative or a shareholder/member company.

Seed production and marketing require special resources and skills. Therefore, individuals and groups identified to do this work should be progressive and prominent in the community, such as leading farmers, innovative village governments, alert businessmen, aggressive agribusiness groups, seed distributors, seed dealers, government employees desiring new opportunities for service, and autonomous government institutions with special resources to contribute. Ability, commitment, and a desire for service to the farming community are extremely important, regardless of who forms the enterprise.

Organizational Patterns

The organization of a good seed enterprise requires that the components of the infrastructure be combined harmoniously at the right place and time. A good seed organization requires the means to grow, dry, process, and store good quality seed that will give better returns to the farmers and the means to market the seed efficiently and at the right time. Either as part of the organization or closely linked to it, there must be plant breeding research to develop varieties and a method to supply information to the farmer about the varieties and how to use them.

Table 1 shows the components of a typical seed enterprise and their functions. Research has been listed as optional since many seed enterprises use the results of research from other organizations. Similarly, the supply of seed stock may come from within the enterprise or elsewhere. Countries normally have several kinds of supporting components that can be beneficial to the seed enterprise. The enterprise may or may not use them, although external quality control on seed sold may be mandatory.

The components and the various supporting components of a seed organization should complement each other. Whatever organizational pattern is chosen, its success will depend upon the persistence, seriousness of purpose, and competence of the personnel and agencies involved. The size of the operation, the personalities involved, and the needs of each enterprise will affect the kind of organizational pattern used (typical organizational diagrams for seed enterprises are shown in Figures 5, 6, and 7). In small enterprises one person must bear responsibility for many functions. In large enterprises several people may be involved in each activity.

Management of the Seed Enterprise

The principal objectives of a seed enterprise are to produce and market good quality seed, win the confidence of seed growers and seed users, and ensure a reasonable return on the capital invested. Like any other organization, seed enterprises need efficient, dynamic, and imaginative leadership to achieve these objectives. Much good information on general management is available, but the seed industry has some special management problems:

- Both production and marketing are seasonal.

- Seed production is not like conventional manufacturing in which most processes are under complete control. Frequent changes of varieties, differing climatic conditions, and varying disease and pest threats force sudden modifications in production technology.
- The production steps follow a definite sequence and need constant surveillance, plus immediate action when problems arise.
- Production is dispersed over large areas and involves many seed growers who vary in experience, production capacity, and capability.
- Seed—the end product—is a living material that must be handled carefully and used before it dies.

Because of these features, managers face an array of unique decisions. The seasonal nature of the industry calls for sound financial arrangements. The management of inventories is complicated by seasonal production and marketing as well as by the living nature of seed. Seed of cereals that loses viability is no longer seed but grain, and not even that if it is treated with toxic chemicals; seed of most vegetable and forage species is worthless if it is not viable. Thus risk is rather high and this complicates credit availability and financial management.

Having seed of the right variety produced in the proper amount and available for sale in a particular area at the right time requires thorough planning and sound judgment. Production patterns often need to be changed suddenly. Advice from the research scientists, the production specialists, and the marketing manager may need to be assimilated and utilized in charting a new direction quickly. Ensuring that the marketing personnel are well enough informed about each variety to be convincing in their relationships with farmers requires that managers be knowledgeable in both agronomy and marketing.

Efficient management requires not only general managerial skill and leadership qualities but also the ability to plan, organize, lead, and control all functions of a seed enterprise. The manager and key persons involved in management need a background in crop research, crop production, seed technology, finance, and marketing. The staff at the management level must have integrity and be able to win the confidence of seed growers, shareholders, seed users, and others on the staff.

The manager must know field activities and be prepared to work frequently outside the office. However, because the usual dispersal of seed operations often demands that decisions be made on the spot, it is essential that the management employ a competent field-oriented staff to whom authority can be delegated.

Organizing seed growing on farms totally managed by the seed enter-

Table 1.
Components of a seed enterprise, their functions, and supporting components

Components	Function	Supporting Components
ADMINISTRATION		
Shareholders/Members Forming the Seed Enterprise	Provide share capital; may produce seed if shareholders/members operate farms	
Governing Body	Determine policies; make important decisions; approve budget; responsible to members or shareholders	
Manager	Chief executive officer to implement policies and decisions of governing body	
Finance	Responsible for initial budget preparation, receipts, payments, and accounts	Banks and other credit sources: provide credit for seed enterprises, seed growers, and farmers using seeds
RESEARCH: PRIMARILY GENETIC RESEARCH (Optional)	Development and testing of varieties plus limited agronomic research	Crop improvement research: responsibility for government, including both genetic and agronomic research
SEED PRODUCTION		
Management	Responsible for all seed production, processing, and storage	
Field Operations	Arrange all seed growing, perhaps including seed stock, guide and supervise seed growers	Seed stock supply: may be supplied from public research program, a special agency, or by a unit of the enterprise

Processing — Drying, cleaning, sizing, treating, and bagging seed

Storage — Bulk and bag storage in production area and at end use area

QUALITY CONTROL

Guidance and Inspection — Responsible for a thorough quality control system to assure buyer of a good quality product

Seed certification services: external control on genetic and physical purity to help assure buyer of good products—voluntarily used by seed enterprise

Guide seed growers and inspect seed at all stages to assure good quality seed production

Sampling and Testing — Sampling from field, processing plant, and storage facility to test and evaluate quality

Marketing Control — Grow out tests of random samples to observe plant uniformity and check on effectiveness of the system

Quality control on seed sold: external control on seed quality, mandatory through legislation, a government responsibility

MARKETING

Responsible for distribution, determining need, accumulating supplies, communication, distribution, and follow-through in the field

Education: responsibility of government to extend appropriate technology to the farmer

Distribution — Allotment and movement of seed to end use area

Advertising and Publicity — Acquainting public with merit of product and encouraging purchase

Sales — Wholesale or retail selling

Follow-through in the Field — Visits in the field before and after sales to know farmers' problems, to deal with complaints, and to gather market intelligence

Figure 5. Organizational pattern for a small seed enterprise

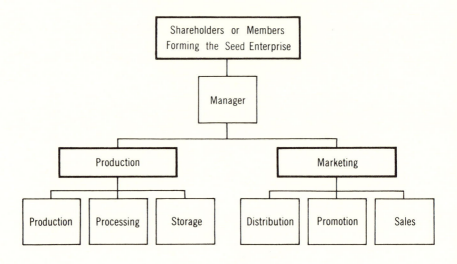

Figure 6. Organizational pattern for a large seed enterprise

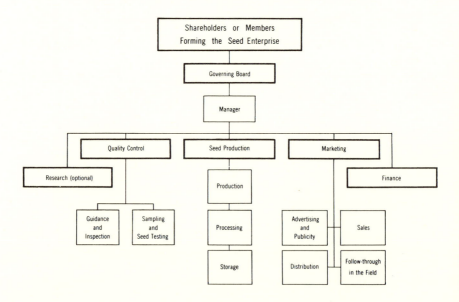

Figure 7. Alternative organizational pattern for a large seed enterprise

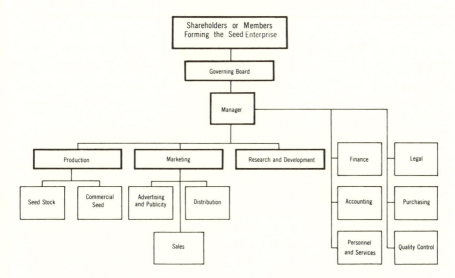

prise or through contracts with seed growers requires certain managerial skills at the farm level if a large volume of good quality seed is to be produced (the next section discusses the seed production requirements more fully). Drying, processing, and storage facilities require a well-trained manager to schedule harvesting and drying operations. Especially in warm humid areas, seed must be dried to a safe moisture content for storage within two to three days of harvest. Thus, seed must be placed in the dryer within hours after it is harvested.

The management of a seed processing plant should focus on ways to achieve effective cleaning, prevent mixtures, maintain lot identity, and avoid damage. These objectives require systematic planning of work; careful supervision of workers to avoid mistakes; thorough knowledge of the seed of the crops involved, their contaminants, and the methods of removing unwanted materials; and a high level of cleanliness and sanitation in the processing and storage areas. In the storage stage, managers must concentrate on good inventory control, systematic testing of seed lots to monitor their viability, and controlling insects and rodents.

An effective quality control program should be in operation. The management must constantly focus the staff's attention on maintaining and improving seed quality. As a reputation for good quality seed is established through good management, seed sales will grow and the monetary success of the enterprise will be assured. The success of a seed enterprise depends upon sound technical and economic decisions, which political decisions should not be permitted to bias or alter.

PRODUCING, PROCESSING, STORING, AND FINANCING THE SEED

While a government must make systematic plans for agricultural development and base the Certified Seed, commercial seed, and seed stock requirements on realistic targets (see Chapter 1), the seed enterprise must also consider its own production plans and special seed requirements to achieve its objectives.

Planning Production

Some seed enterprises start as modest, soundly planned, well-supported projects and grow into large operations. Others begin as complex, high-volume operations and require exceptional managerial skill. The seed demand, availability of skilled manpower, and resources of the enterprise are the major determinants of the appropriate size of the initial operations. The training of personnel should receive high priority to ensure that the enterprise can develop rapidly and soundly.

The kind and quantity of seed to be produced must be planned several years ahead because at least three or four seasons of seed multiplication are usually required to produce Certified Seed or commercial seed from seed supplied by breeders. Nationwide replenishment factors (see Chapter 1) will affect a seed enterprise's plans for the amount of seed it needs to produce of any particular variety. A multiplication rate or factor can be calculated from the estimated yield of seed and the crop's normal seeding rate per unit of area. From projected demand and the multiplication factor, it is possible to determine the number of multiplications that are to be used and to establish the area and the amount of seed needed for each generation. Extra production is needed to cover losses in the field and reductions in the quantity of usable seed that occur during seed processing.

Careful planning is necessary to avoid overproduction or underproduction at each stage of multiplication and to coordinate the agencies or units that maintain and increase seed stock and commercial seed.

Areas for Seed Production

Suitable seed growing areas must be identified. The areas should be capable of reliable, economic production with a minimum of natural hazards. They should have the right soil and climatic conditions for the crops being produced, low weed and disease incidences, assured water and power supplies, ready access by road, and a good communications system.

A point often overlooked is that the area where the seed is used may not be suitable for producing high quality seed; this is especially true for many forage and vegetable species. Seed of these crops is often shipped long distances from seed producing areas to seed consuming areas.

The land for seed production could belong to an enterprise's shareholders, other farmers, the government, or other institutions. The seed enterprise could contract for seed production with any of these groups. When private farmers are involved, they should be progressive, be located near the seed drying and processing plant, have operational farm holdings of an efficient size, and have the needed equipment. The contract should be fair to both the grower and the seed enterprise and clearly specify the responsibilities and duties of both parties. Nevertheless a contract is only a means to an end—it has to be based upon a mutual respect between the seed grower and the seed enterprise (Appendix C lists points often included in a seed contract).

Efficient Cultivation

Seed production requires good cultivation practices to maximize the multiplication ratio. Seed growers need fertilizers, pesticides, and herbicides along with seed stock, and it may be necessary for a seed enterprise to arrange to supply these needs. Cooperative programs among seed growers have proved beneficial in handling some operations. Custom services could be provided to farmers by the seed enterprise or by another agency in the area.

Maintaining Trueness-to-Variety

Multiplying a crop so that it continues to be true to the variety requires care in selecting the field, in planting, in removing plants that are not true to the variety, in adequately isolating the field, and in harvesting. It may be difficult to maintain a proper isolation distance around seed fields if individual holdings are small. Developing seed growing around selected villages where the growers agree to locate seed fields in a compact area is one solution. Or, the pollinator or other noncontaminating seed could be supplied to the neighbors of the seed growers. As a last resort, legislation may be required that prohibits the deliberate planting of varieties that create isolation problems.

Timely Harvesting and Efficient Drying

Harvesting at the right stage of seed maturity is essential to achieve a maximum yield; to minimize field deterioration, mechanical injury, and damage by birds, rodents, or molds; and to obtain high quality seed. Seed health and vigor peak at the time of physiological maturity and then start to deteriorate. Prompt harvesting with properly cleaned and adjusted equipment is more critical with seed than with grain. Proper drying is necessary to maintain seed viability. If climatic conditions are unfavorable, seed may have to be dried artificially and provision needs to be made for this possibility. The arrival of seed at the drying and processing facility must be organized so it can be dried immediately after it is harvested.

Guidance and Training for the Seed Growers

Seed growers need to be supplied with recent and concise information on how to grow and harvest good quality seed. This information should be based on local experience and on sound research concerning seed production technology. Guidance can also be obtained from programs in developed nations, but their procedures often need adaptation to local conditions. Useful sources of reference material on seed production include the institutions involved in training cited in Appendix G; several of the organizations, associations, and international research centers listed in Appendix H; and national crop research programs. Also, several of the references listed in the Bibliography include information on seed production.

Seed growers can benefit highly from visits to research stations, and new growers can benefit from visiting experienced growers in their own country or abroad. As production requirements change, seed growers need continuous guidance. The field staffs of seed enterprises and other supporting agencies must be trained to provide effective advice.

Encouraging the Grower to Produce Better Seed

Because seed growers vary in background, skills, education, adaptability, and integrity, managers of seed enterprises must provide incentives for the growers to produce better quality seed. The first step is to make sure a grower thoroughly understands the importance of seed quality. As a grower produces better quality seed, his yield and his returns will be higher. In addition to supplying growers with literature on seed, the seed enterprise could organize short training courses using successful seed growers and scientists from research institutes and agricultural universities as instructors.

Seed quality and production will rise as each seed grower gains experience. The knowledge that he can continue in the program if he does well will motivate a grower to improve production. Higher prices should be paid for better quality seed. If germination is a problem, growers could be paid on a graded scale based on the percentage of germination above a specified minimum for each seed lot. To do so, a skilled seed testing technologist familiar with the germination characteristics of the crop would be needed. A graded scale could also be based on the seed's physical purity, grain character, weed seed percentage, and other quality factors. Seed samples from each grower should be grown out periodically to evaluate how thoroughly such tasks as detasseling and roguing have been done, and the care taken in harvesting the crop. The fact that such "grow-out" tests are conducted will encourage the seed grower to do a better job.

Seed Processing and Storage

At the processing plant, inert material and alien seed are removed to improve the appearance, planting quality, and acceptability of crop seed. Accurate sizing of seed of some crops and the removal of lightweight seed improve the planting value of the seed. For many crops, treating the seed with a fungicide or an insecticide, or both, helps protect seed until it has germinated. Suppliers of seed treating materials and many reference works can provide current information on which chemicals to use.

The managers of a seed enterprise need to be sure that equipment is being maintained and used properly. Attempting to process more seed than the equipment can adequately handle causes poor grading and poor cleaning. On the other hand, not using equipment to full capacity during the season is inefficient. The training, experience, and judgment of the equipment operator are critical. (See Chapter 8 for further information on seed drying and processing.)

Throughout the entire processing and storage period, managers must be sure that the staff gives adequate attention to seed quality by making timely checks of all seed lots and keeping records that provide up-to-date information about the seed lots (examples of forms for record keeping are given in Appendix C).

Although maintaining the viability of stored seed until planting can be the difference between profit and loss, new seed enterprises seldom have adequate space. A seed enterprise needs adequate storage facilities for the commercial seed that will be planted during the coming season, for the reserve commercial seed and unsold commercial seed that will be kept for planting in the next season, and for the seed stock being held for multiplication in subsequent seasons. The requirements for storing seed to be planted during the current season are less exacting than those for reserve and unsold commercial seed or seed stocks. Managers must recognize these differing needs and strive to minimize losses from seed deterioration. Many publications present the detailed requirements for storing seed (more details on seed storage facilities are included in Chapter 8).

Finances

As discussed earlier, the availability of requisite finances for seed enterprises and seed growers must be assured. Cultivation of a seed crop requires a high investment in inputs as well as investment in land development and field machinery.

ASSISTANCE FROM FOREIGN SEED ENTERPRISES

External assistance in building seed supplies may come from various sources. However, foreign seed enterprises can make special contributions to seed enterprises in a developing country. In recent years, several seed companies in North America and Western Europe have developed international programs. Before considering how foreign seed enterprises can be most useful, it may be helpful to review their main characteristics.

Because seed companies are relatively small in comparison with major corporations, the capital available for foreign investment may be limited. Usually seed companies are headed by capable managers, and their personnel have practical experience in crop production and seed technology. But these companies do not have managerial and technical personnel who can be made available for prolonged consultation or the time-consuming activities required to start a new operation.

International seed companies possess a wide range of germplasm. From this germplasm, their research programs have developed many varieties that are unavailable elsewhere. The companies depend on the use of these varieties for sales in highly competitive world markets. When varieties are adapted for use in developing countries, the originating company will want its proprietary rights recognized and protected. They, like comparable local enterprises, usually expect a fee for the use of such varieties.

Foreign seed companies might participate in the development of a new

---

Types of Assistance Possible from a Foreign Seed Enterprise

Distribution network
  Seed supplies

Seed franchises
  Germplasm
  Technical aid
  Managerial assistance

Consultants
  Advice on establishing seed enterprises and production and
    marketing activities

Direct foreign investment
  Capital
  Research and development (resulting in new varieties)
  Management and marketing skills improved
  Greater export opportunities

---

seed industry in several ways: (1) through their distribution network, (2) by the use of seed franchises, (3) by acting as consultants, and (4) through direct foreign investment.

Distribution Network

In countries where the market for seed of a particular crop is small, where the seed is difficult to produce, or where no seed industry exists, foreign companies usually appoint local distributors to import and sell seed. The value of this type of relationship is often overlooked. The role of imported seed as part of the overall supply of a nation's needs was discussed earlier in this chapter. A local distributor can function as an information specialist for the crop seed he stocks. He is in close contact with his supplier from whom he receives technical information, literature, films, and other materials. Some distributors visit the company's research departments, and the company's technical representatives visit the distributor to see local farming conditions. More significant than being merely a supplier, the distributor frequently becomes a "seedsman" who produces and markets seed that previously was imported.

Seed Franchises

A franchised seed operation is another way foreign enterprises can contribute to a country's seed program. Through a franchise, a local seed enterprise will receive germplasm, technical aid, and, if necessary, managerial assistance. The terms of each franchise agreement vary, but in general all the resources of the international seed company, except capital, are available to the local company. For a franchise to work, proprietary rights must be recognized and an arrangement reached under which the originator receives payment for the use of its varieties.

Although a franchise is for the production of seed of varieties developed by the foreign company, a franchise holder might also produce and market seed of local varieties. On a national basis, through several franchised seed operations, a wide range of seed of both proprietary and public origin could be produced and marketed.

Consultants

Countries establishing seed enterprises, starting seed production and marketing activities for certain crops, or revitalizing their present program might hire consultants from foreign seed enterprises.

Direct Foreign Investment

Before making an investment overseas, a company evaluates the potential size of the market and assesses the probability of realizing an adequate return on its investment. Investors attempt to select situations that appear most favorable. If foreign investments are welcomed in a country, administrators responsible for the development of a seed industry need to consider the extent to which foreign seed companies should be encouraged. The interest of foreign seed companies can be stimulated through provisions for the return of adequate profits to the foreign company, opportunities to import equipment and seed stocks without long bureaucratic delays, a cooperative attitude toward private crop-breeding research and the introduction of varieties developed, encouragement of marketing by seed enterprises and marketing groups without direct or indirect government price fixing, and seed quality control measures that are not unrealistic.

While the benefits of foreign seed technologies, germplasm, and seed of named varieties can be made available through franchised operations and distribution outlets, direct foreign investment can further development in other ways. Joint research activities can increase the flow of better varieties for farmers. The management of seed enterprises can be improved. Financial capability can be strengthened with new capital, and possibilities for loans may occur. Additional marketing skills can be introduced to the seed enterprises. Opportunities for exporting seed may arise through reciprocal arrangements with the parent company. And local personnel can be trained by the foreign company.

Although too much dependence on foreign technology can be a disadvantage, most countries will also be developing their own technology and encouraging seed enterprises that have no foreign participation. The foreign exchange used to return profits to the parent company can also be a disadvantage; however, this must be balanced against the possible reduction in the cost of imported seed and the added benefits to the local economy from use of the varieties developed.

Many jointly owned companies that have foreign participation exist in developed countries, and examples can be found in a few areas of the developing world as well. Administrators should evaluate successful examples in other countries if none exists in their own country.

Most international seed companies are willing to consider arrangements such as distributorships, franchises, and various levels of equity participation. These companies can help the development of a strong local seed industry. The Fédération Internationale du Commerce des Semences and the Industry Council for Development (Appendix E) can provide information about companies that have an international interest in seed.

Building the Seed Supply: Major Policy Points

1. Whether seed importation is to be encouraged or discouraged, the crops involved, and the restrictions, if any, that are to be applied
2. Whether local seed production is or is not to be encouraged
3. If local seed production is encouraged, whether seed enterprises are to be formed, the manner in which this is to be done, and whether private investment is to be stimulated
4. Kind of assistance that can be provided to stimulate the formation of seed enterprises and hasten their growth
5. Nature and amount of involvement expected from foreign seed enterprises

REFERENCES

Delouche, J. C. 1969. *Problems and Prospects in Seed Program/Industry Development in the Less Developed Countries.* Mississippi State: Mississippi State University.

Douglas, J. E. 1973. *Seed Production, Technology, and Industry Development in India—Final Report.* New Delhi: Rockefeller Foundation.

Feistritzer, W. P., ed. 1975. *Cereal Seed Technology.* Rome: FAO.

Government of India. 1968. *Seed Review Team Report.* New Delhi: Department of Publications.

Law, A. G.; Gregg, B. R.; Young, P. B.; and Chetty, P. R. 1971. *Seed Marketing.* New Delhi: Mississippi State University, National Seeds Corporation, and U.S. Agency for International Development.

Tarai Development Corporation, Ltd. 1975. *Tarai Seeds Development Project—Organisation and Operation.* Pantnager, India: G. B. Pant University of Agriculture and Technology and Tarai Development Corporation.

5
Seed Quality Control

Both the seller of seed and the government bear a special responsibility for the quality of seed sold because a satisfactory harvest frequently hinges on how good or how bad the farmers' planting seed was. Administrators often ask, Can we solve our seed problems with a good seed certification program? How much can a seed testing laboratory really do about improving seed quality? When should we do something about seed legislation? How can the operation of existing programs be improved to raise seed quality?

Seed quality can be gradually improved if the people in a seed program appreciate its importance, if a realistic seed certification system is functioning, if a seed testing laboratory is evaluating large numbers of seed lots from production and marketing programs, if well-conceived legislation is applied uniformly, and if each segment of a program is properly organized and managed.

IMPORTANCE OF GOOD SEED QUALITY

While almost everyone agrees that planting good quality seed is important, definitions of good quality differ. Many farmers judge seed quality by how the seed looks—plumpness, color, and absence of extraneous materials. They want seed of the right crop and, often, of a particular variety. Although farmers can usually identify the kind of seed, they are seldom able to identify the variety, nor can they determine how well the seed will germinate by looking at it.

A seed technologist assesses seed quality more precisely. Good quality seed is thought of in terms of high analytical purity (low content of inert matter and seed of weeds or other crops), high germination percentage, and freedom from seed-borne diseases. The seed must also be true to kind and variety. It is expected to be of an improved variety that will provide good results under the conditions for which it is intended. In other words, a seed technologist usually expects the term "improved seed" to mean good quality seed of an improved variety.

Seed Quality and Successful Programs

Neither good quality seed of poor varieties nor poor quality seed of superior varieties serves farmers well. Crop research and development programs must be concerned not only with the performance of varieties, but also with the quality of seed available to farmers. To be successful, seed programs must consistently provide the farmer with better quality seed than he can produce himself.

Ensuring Good Seed Quality

The obligation of administrators and leaders at all levels of a seed program to assure farmers of good seed quality of improved varieties is complicated by the inability of any one person or organization to control the quality of all seed used. Steps to create good seed quality must stretch from the crop research and development program through the initial seed multiplications and to all subsequent production, drying, processing, and distribution activities:

- in the production stage—proper fertilization, adequate water, sufficient isolation, adequate roguing, timely harvest, and care in harvest;
- during drying—timeliness and correct temperatures;
- during processing—care to increase pure seed percentage, to avoid admixtures, to minimize damage to the seed, to provide the proper seed treatment, if necessary, and to put the seed in a satisfactory package at a safe moisture level;
- during storage—proper seed lot identification and suitable conditions to avoid rapid loss in germination;
- during distribution—care in transporation and storage to avoid excess humidity or heat, to prevent contamination, and to maintain proper identity of the seed lot until it is sold.

Maintaining good seed quality through all these steps requires knowledge of what to do and dedication to high seed quality. The seed enterprises are the first line of defense against poor quality. Their internal quality control programs should provide the means to identify a seed lot as it moves from the seed grower to the point of sale. Such a lot system establishes the accountability for seed necessary for effective seed certification and law enforcement programs.

In large seed enterprises, quality control specialists make sure at each step from production to distribution that all technical and administrative details are carried out thoroughly and on time. In small enterprises, managers may have to assume this responsibility.

Ways Governments Can Promote Good Seed Quality

Various government activities and programs can increase awareness of the desirability of good seed quality. When government resources are used to train seed technologists, whether inside or outside the public sector, the courses should emphasize ways to achieve good seed quality. Educational programs for farmers can raise their concern about the quality of the seed they buy or save for planting. A government's crop research and development programs can demonstrate the importance of seed quality to farmers as a part of field trials (Chapter 6 under "Linking Research and Practice") and, by example, in the production of Breeder Seed and Basic Seed. Even before a complete seed program is operating, a government can promote good seed by establishing quality standards and ensuring, through administrative orders, that its agencies distribute only seed that meets the standards set.

When a government buys seed, domestically or abroad, price should not be the sole criterion. Before a purchase is made, the varietal purity, analytical purity, germination level, moisture level, and kinds, if any, of seed-borne diseases present should be fully evaluated. In particular, the value of different seed lots should be compared on the basis of cost per unit of pure-live seed (percentage of pure seed that germinates). That comparison often reveals that seed with the lowest price is the poorest buy. Paying slightly more may result in substantially higher yields.

The foregoing actions complement formal government seed quality activities—certification, testing, and legislation. These steps can contribute to the stable and systematic development of a program with an emphasis on quality.

SEED CERTIFICATION

A seed certification program is one tool for producing genetically pure, good quality seed of improved varieties. In most developing countries, seed certification means that certain quality requirements are fulfilled and made evident for the buyer. Seed quality is usually appraised by a seed lot's trueness-to-variety, physical purity, content of other seed (certain weed species may be prohibited), germinability, seed health, and moisture content. However, seed certification under the certification schemes of the OECD (Organization for Economic Cooperation and Development) and some agencies in North America is based wholly on trueness-to-variety; that is, lots are certified true to the variety's characteristics, including variations, as described by the breeder (trueness-to-variety, however, does not necessarily mean *extreme* uniformity). Good evidence of stability in a variety's composition and performance is also expected.

In many countries the quality level of Certified Seed is officially recognized and controlled. The official certification label, which appears on every Certified Seed package, contains data that allow the farmer to have confidence in the material planted. To deserve the farmer's confidence, a seed certification system must be totally independent of seed production and marketing programs.

A certification system normally deals only with varieties that are at least equal to those previously released, but factors such as disease resistance, quality, and maturity must be considered in addition to yield. Seed of these varieties is multiplied through a series of generations (Breeder Seed, Basic Seed, and a class of Certified Seed). The source of the seed planted must be verified at each step. Field inspections, laboratory tests, and, often, plot tests to check varietal purity of individual lots combine with the eligibility of the variety and a verified seed source to make a total system.

But seed certification is not a seed program in itself, nor can it be a substitute for the many other elements discussed in this book. Groups and individuals still must be organized to do the planting, roguing, harvesting, processing, storage, and marketing. Figure 8 illustrates the relationships that must exist between seed certification and production and marketing activities, though they are independent of one another. The seed certifying authority can be a catalyst of these activities. Without this parallel development, a seed certification program has no purpose.

At the start, a certification program should be voluntary, and it should concentrate on the most important crop species. A program can expand gradually as the demand for good quality seed of improved varieties grows. Later, mandatory seed certification can be considered on its own merits. In many advanced programs, however, certification continues to be voluntary.

The sale of Certified Seed can help a new seed enterprise establish its reputation as a producer and distributor of good quality seed and reduce disputes about varietal purity. For the seed industry in general, seed certification forms a basis for quality differentiation, especially in the early stages of a program.

Figure 8. Seed certification is linked with production and marketing

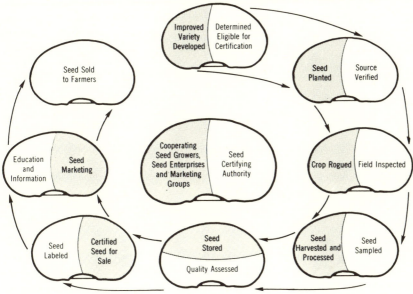

Terminology

The terms Breeder Seed, Basic Seed, and Certified Seed are used in this book to identify different generations and classes in a seed certification program. The seed quality requirements in these classes differ, especially with respect to varietal purity percentage. Breeder Seed and Basic Seed are intended for seed multiplication only. The generations after Basic Seed are called Certified Seed and may be identified as Certified Seed–1, Certified Seed–2, and so forth.

The terms used to identify the generations may differ among countries. The OECD seed certification schemes for seed moving in international trade use certain terms (see Appendix E). The Association of Official Seed Certifying Agencies (AOSCA) has standardized terms for the United States and Canada. Table 2 compares the terminologies of the OECD and the AOSCA with the terms used in this book.

New seed certification systems or programs that are being changed should use terminology that is as similar as possible to current usage. For countries expecting to participate in the OECD schemes, the use of the OECD terms will minimize confusion.

Table 2.
Comparison of seed certification terminology used in this book with that of two international seed organizations

Responsibility and Use	OECD[a]	AOSCA[b]	This Book
Multiplied by plant breeder for multiplication (may include more than one generation)	Pre-basic Seed	Breeder Seed	Breeder Seed
Multiplied under the plant breeder's care or by a special agency	Basic Seed	Foundation Seed	Basic Seed
Multiplied by seed growers, seed farms, or seed enterprises for multiplication and sale or further multiplication	Certified Seed, first generation	Registered Seed	Certified Seed-1
Multiplied by seed growers, seed farms, or seed enterprises for sale commercially or further multiplication	Certified Seed, second generation	Certified Seed	Certified Seed-2
Multiplied by seed growers, seed farms, and seed enterprises for commercial use	Certified Seed, third and subsequent generations depending on the variety involved	Certified Seed, third and subsequent generations used with some crops	Certified Seed-3 and subsequent generations; considered desirable with self-pollinated crops in many new programs

[a] OECD: Organization for Economic Cooperation and Development
[b] AOSCA: Association of Official Seed Certifying Agencies

Steps in Certifying Seed

Seed certification is a system that incorporates certain basic steps: determining the eligibility of varieties, verification of seed source, field inspection, sampling, seed testing against quality standards, labeling, conducting variety control plots, and education and information.

Determining Eligibility of Varieties

New varieties come from national plant-breeding activities, other local sources, or sources outside a country. The varieties should be tested in different regions of a country to evaluate their yield and quality characteristics under many growing conditions. A variety review and release committee (see Chapter 2 under "Relating Test Results to Variety Release") should decide which varieties to certify. The names of these varieties should be published annually, and they should be well publicized.

Verification of Seed Source

Although the acceptability of a certifiable source of the seed planted must be verified for each generation of multiplication, there is no need to limit the number of generations of Certified Seed as long as varietal purity is maintained and the seed source is documented through certification. Increasing the number of Certified Seed generations and, hence, the volume of Certified Seed produced, reduces the need to produce as much Breeder and Basic Seed. However, in cross-pollinating or open-pollinating species, biological reasons may limit the number of generations. Additional generations are possible only when the seed source can be verified and the multiplication chain is maintained.

Field Inspection

The first stages of certification take place when seed certification technologists check the seed growers' fields. They verify that the seed planted is eligible to produce Certified Seed and note the previous crop in the rotation. An inspection is made to assess off-type plants, other varieties, weeds, other crop plants, and diseases present in the field. The distance to other fields is checked, particularly in the case of a cross-pollinating species, to avoid unwanted crossing and other contamination.

Hybrid seed production fields are carefully monitored for unwanted pollen sources. In maize, for example, detasseling is specifically checked in order to maximize the hybrid effect in the following generation intended for planting. The OECD has published guides for the field inspection of cereal and forage (herbage) seed species.

Sampling

In most countries the processed seed is officially sampled by seed certification technologists. The seed testing laboratory tests the samples before certification and labeling. The International Seed Testing Association (ISTA) has established rules that specify recommended methods of sampling, seed lot size, and sampling intensity.

Seed Testing Against Quality Standards

Seed testing to determine the different elements of seed quality is a necessary part of a certification program. At the start, minimum seed standards for the important quality elements should be related to the available quality. Setting standards too high could stifle the whole program. But when progress is made, standards should be raised. Minimum seed standards should be based on the data from the seed testing laboratory.

The standards should not be inflexible. In some years, the weather or other conditions may result in a need to modify the standards temporarily to avoid a seed shortage. Such a decision should be made cautiously to ensure that respect for the seed certification system is maintained. Varietal purity of seed, however, is not affected by seasonal or climatic conditions so those standards are not normally changed.

Labeling

When a seed lot meets the minimum standards for a certain quality class, certification labels are put on every seed container. The certification label is the document that shows the standards have been met.

Conducting Variety Control Plots

The progress made by plant breeders in developing a better variety can be lost quickly if admixture occurs during the various stages of multiplication. Because some varieties are difficult to identify during field inspection, many advanced seed programs operate special field plots to check the genetic purity of seed lots that have been certified. Such field plots are also used in assessing and approving seed for the next generation of seed production. Samples are drawn from seed lots and sown in a control field for comparison with plants grown from the original Breeder Seed. The OECD and ISTA have guidelines and rules for plot testing various species. In new programs, plot testing can be used if trained personnel and suitable facilities are available.

Education and Information

Convincing farmers of the importance of using Certified Seed of suitable varieties and continuing to inform them of that importance are necessary parts of seed certification. The field staff, too, needs to be "seed educated" in order to help farmers understand the importance of good quality seed (see Chapter 6).

Sources of Technical Details

Guidelines for procedures and methods related to seed certification can be found in the FAO's *Cereal Seed Technology*; the OECD schemes; the suggestions of the International Union for the Protection of New Varieties of Plants for describing crop varieties (see Appendix E); ISTA's special publications on varietal purity and on variety testing; and the *Certification Handbook* of the Association of Official Seed Certifying Agencies. Many countries have reports and publications on seed certification programs as well.

SEED TESTING

Quality control programs are pointless unless they involve seed testing. Conversely, a seed testing laboratory has little value unless it is a part of a seed certification program, a seed law enforcement program, or a production and marketing activity (Figure 9 illustrates the users of seed testing results).

Seed tests can provide information on pure seed, other crop and weed seed (by percentage and number per unit weight of different species), inert matter, normal and abnormal seedlings, fresh or hard seed, dead seed, and moisture content. Certain countries that have special staffs and facilities are able to test varietal purity—in the field, greenhouse, or laboratory—seed health, and seedling vigor. Testing cannot make the seed better than it is, but, based on test results, advice can be given on how to avoid or remedy poor seed quality.

Applied Seed Testing

Testing is the final step in judging a seed lot's eligibility for certification. Through testing, the quality characters, as specified in the certification standards, are assessed. Seed testing is also essential for enforcing legislation that specifies the quality requirements for seed sold. When

Figure 9. Users of seed testing results

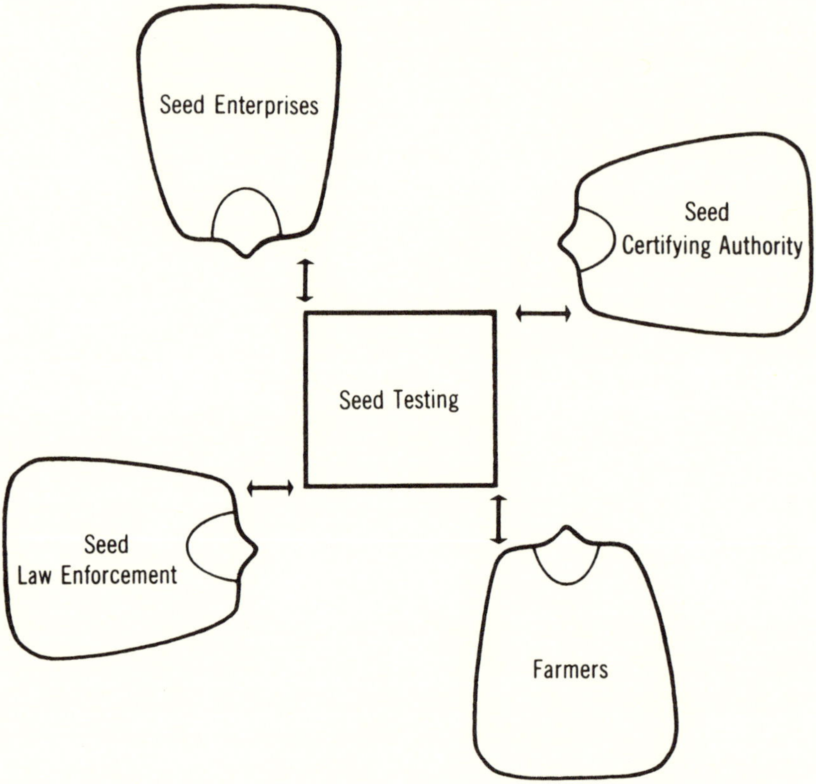

seed is being marketed, official samples are taken and sent for testing.

Seed enterprises must use testing every day to control the quality of the seed at all stages—from receiving uncleaned seed through the drying, processing, storage, and distribution phases. Uncleaned seed is commonly tested to establish the basis for payment to the grower. The laboratories that do this may be official or a part of the enterprise. Similarly, marketing groups need information on seed they purchase and sell.

Service testing for farmers who use their own seed will enable them to identify poor quality seed lots. Farmers can be encouraged to send samples to a seed laboratory for analysis if testing is free or inexpensive. It is cheaper for a government to subsidize such tests than to deal with food shortages resulting from poor seed.

Because seed is a living material affected by many external conditions, problems can occur at any stage of the seed multiplication and distribution program. Seed technologists are among the first to notice potential threats to seed production. Therefore, at least the main official laboratory of a country should be encouraged to undertake practical research. In addition, in the many countries that lack seed testing rules for certain species, suitable procedures need to be found. The seed testing staff members should also help establish appropriate seed certification standards based on testing. They need time to study the literature on new techniques and to conduct other research relevant to local conditions.

Finally, a laboratory staff should emphasize good seed quality and teach proper sampling to seed growers and other seed technologists because the correctness of testing results can be no better than the accuracy with which the seed sample was collected. To establish a broad appreciation of good quality seed, a seed testing laboratory should have an open-door policy, and its staff should actively disseminate information outside the laboratory to producers and users of seed. A laboratory should never lose contact with the community. The ultimate administrative objectives in seed testing should be accurate and prompt analysis of seed and reporting of results.

International Procedures

The International Seed Testing Association has established internationally used seed testing rules and guidelines. ISTA has also made provisions for reporting quality on special certificates, primarily in connection with the export and import of seed. The rules are mandatory only when issuing ISTA certificates. An ISTA certificate is just a declaration of test results and makes no judgment of quality—evaluation is up to the buyer. A government can make an application for ISTA membership when the seed testing laboratory is fully staffed and equipped and after it is ready to be designated by the national government (details about ISTA are given in Appendix E).

SEED LEGISLATION

Is seed legislation necessary? A few countries have no seed legislation but produce high quality seed, and some countries have comprehensive seed legislation but produce low quality seed. Until recently, two of the world's largest producers of high quality seed had no seed legislation as such. They controlled seed quality through a certification scheme, education, and general statutes.

Each government must ask, Where are we today? What is our prob-

lem? Can it be solved by special seed legislation or must some other action take place before seed legislation can be of value? What kind of
legislation is needed? Are the preconditions for foreign loans the only
reason for legislation? Can legislation to control seed quality wait until
there are seed dealers to control or until it is needed?

When to Legislate

It is best not to legislate until it is needed—and not to adopt more
legislation than is necessary. Seed legislation in developing countries can
be justified when the emphasis is making everyone concerned with seed
production and marketing aware of the importance of seed quality, when
it stabilizes quality standards at practical levels, when it provides continuity to the development of a program, when it facilitates the establishment of reputable seed enterprises and marketing groups, and, ultimately, when it increases the availability of good quality seed. If legislation
cannot promote these objectives, it should be postponed.

Types of Seed Legislation

Seed laws can be adopted for establishing (1) crop research and evaluation systems, (2) seed certification programs, (3) marketing requirements
for categories of seed, including imports or exports, (4) seed testing
responsibilities, (5) plant variety protection or a breeders' rights system,
or (6) a plant quarantine program. But in the logical development of an
integrated seed program, legislation may be one of the last steps to consider. For example, a requirement for the labeling of seed offered for sale
is senseless if there are no seed enterprises or other sellers of seed. It is
similarly futile to enact labeling legislation if sellers cannot determine the
quality of seed offered for sale because no seed quality laboratories exist.

When a crop research and evaluation program, a seed certification
program, and seed enterprises exist, a broad spectrum of marketing
legislation can be considered. At one extreme is legislation that requires
total "pre-marketing control," or control before seed is on the market
(Figure 10). Under this type of legislation, every variety must be officially approved for planting prior to sale, and all seed sold must be certified for genetic purity and must meet minimum quality standards.

At the other extreme, legislation may require only that seed be labeled
with certain information that is truthful. The buyer chooses whatever
variety and quality of seed he desires. This is quality control only when
seed is on the market, or marketing control.

In a marketing control system, seed lots offered for sale might be

Figure 10. Quality control systems

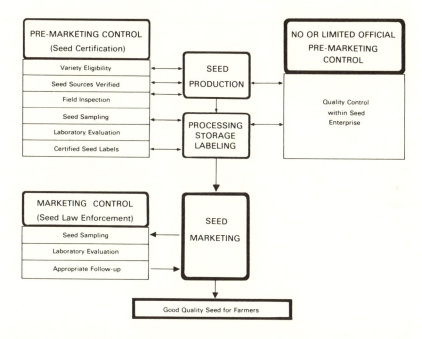

sampled randomly by a seed law enforcement technologist assigned to visit sellers regularly when seed for planting is moving to the buyer. Seed enterprises can also be required to notify seed law enforcement technologists when seed is ready for sale. If this is done, every lot of seed is *available* for sampling before sale, but the enterprise is not required to withhold a lot from sale until it is sampled. The samples collected are tested to determine whether they are truthfully labeled. If they are not, the remaining seed is removed from sale and the labeler is held responsible under the law.

Under a pre-marketing control system, seed of poor adaptability and poor quality is less likely to reach the grower. However, the cost of and possible delay in officially evaluating new varieties before they can be marketed are disadvantages. Under a marketing control system, the advantages and disadvantages are just the reverse.

In Europe, pre-marketing control is prevalent for some kinds of seed whereas in the United States marketing control plus voluntary seed certification (voluntary pre-marketing control) is used. Various combinations are in use in other countries. For example, certification may be voluntary or mandatory; a marketing control system might cover all or only some kinds of seed and have different degrees of mandatory provisions.

Which policy to pursue depends upon the type of government, the educational level of the nation, the resources available, whether seed enterprises exist, and how much improvement in crop production is desirable or possible through better seed. Many seed programs start a voluntary pre-marketing control system by certifying seed of approved, superior varieties grown by farmers and seed enterprises. This approach does not interfere with other seed supplies and permits the entire program to improve gradually. Later, a marketing control system can be added to raise the quality of seed not included in the certification system.

Whether to have a pre-marketing control system or a marketing control system must be determined before a decision can be reached on what kind of legislation should be enacted. Government administrators and leaders must answer these questions for themselves, taking into consideration local needs and the fact that seed legislation is not just to protect farmers but to promote fair competition among seed enterprises and other seed sellers as well. Excessive regulation stifles the industry.

Drafting Seed Legislation

Once broad principles have been decided upon, seed legislation can be drafted for discussion with public and private groups. The best way to do this is to form a committee of representatives of different interests to advise the appropriate government agency during the drafting and reviewing stages. A seed law may contain any of the following basic sections.

Definitions. Definitions should be prepared after the rest of the draft has been basically agreed upon. Otherwise, some terms will be omitted, or terms that do not appear in the final draft will be defined unnecessarily. Definitions should be the first section of the final law.

Crops covered. This section should specify which crops or types of crops the law will cover. Alternatively, authority can be delegated to the minister of agriculture to specify the crops to be covered by regulation or decree. The latter approach is more flexible and allows for gradual coverage, crop by crop, as determined by the progress of the seed industry and the need for control.

Normally food crops and other major crops are included first. Whether to include crops reproduced vegetatively is an important question. Special clauses will be needed for vegetatively reproduced crops because the steps for sampling and evaluating quality are different from those for crops reproduced by seed. However, if vegetatively reproduced crops are economically important and the quality and identity of the material sold can be improved by legislation, such crops should be included.

Harmful weeds. This section should list the most harmful weeds *disseminated in seed* and establish prohibitions or limitations on their presence in crop seed. The list should not include weeds that are not normally found in sowing seed or weeds that are easily controlled in the field. Alternatively, the responsibility for naming harmful weeds and placing a limitation on their presence can be given to the minister of agriculture or his agent. This approach provides the flexibility to adjust to new developments in weed control and changes in the threatened introduction of dangerous weeds.

Labeling. The extent to which seed should be labeled when placed on the market should be included in this section. The language or languages to use for labeling should be specified. Labeling needs will vary with the system of control adopted.

If seed is certified with minimum genetic purity and quality factors are officially tested before sale, the labeling might only indicate the class of Certified Seed, the name of the crop, the variety, the lot identification, and the name of the certifying authority. Such a label (Figure 11) might also be required to show what the minimum standards are, with or without the actual percentages found by testing; whether the seed has been treated; and the date beyond which the label is no longer valid. The producer of the Certified Seed should also be identified by name or code number.

Under a marketing control system, an analysis label (Figure 12) is required on all seed sold. The most complete labeling would include:

1. Name of the crop and variety
2. Lot identification
3. Origin
4. Percentage of pure seed
5. Percentage and/or number per unit of all weed seed
6. Name and number per unit of harmful weed seed
7. Percentage and/or number per unit of other crop seed
8. Percentage of inert matter
9. Percentage of germination
10. Percentage of hard seed
11. Date of the germination test
12. Whether seed has been inoculated with rhizobium
13. Whether the seed has been treated and the name of the toxic

Figure 11. A Certified Seed label containing only information essential for seed certification. An analysis label would be desirable in addition or might be required by a Seed Act.

CERTIFIED SEED

Certified by: National Seed Certifying Authority

CLASS OF SEED: _____

KIND: _____

VARIETY _____

LOT No. _____

TAG VALID FOR (YEAR): _____

CERTIFIED SEED PRODUCER CODE No. _____

NAME: _____

ADDRESS: _____

The seed to which this label is attached is equal to or above the minimum requirements for this class of certified seed as specified in the Certified Seed Regulations and for the requirements as given in Seed Regulations.

NET CONTENT: _____ KGS.

Figure 12. An analysis label which includes details often required by a Seed Act

ANALYSIS LABEL

NAME: _____

ADDRESS: _____

KIND: _____

VARIETY: _____

LOT No. _____

GERMINATION (MIN.) ____ %   INERT MATTER (MAX.) ____ %

HARD SEED (MAX.) ____ %   OTHER CROPS SEED (MAX.) ____ %

PURITY (MIN.)   ____ %   WEED SEED (MAX.) ____ %

DATE OF TEST: _____

ORIGIN: _____

POISON, TREATED WITH _____

(DO NOT USE FOR FOOD, FEED OR OIL PURPOSES)

NET CONTENT: _____ KGS.

substance together with any warning necessary concerning the handling of the seed
14. Net seed content
15. Name and address of the labeler or seller

If legislation for both seed certification and marketing control is developed, the labeling requirements must reflect the needs of both systems. Thus, the Certified Seed label (Figure 11) and analysis label (Figure 12) would both be utilized. Or the information from both of these labels could be combined in a single label (see Figure 13).

When a marketing control system is used and more protection of the consumer is desired by a government, minimum quality standards for all kinds of seed—or for those of greatest economic importance—can be established for purity, germination, content of moisture and weed seed, or health condition. A simplified labeling system then may be used on some or all seed giving only the name of the crop and variety, the lot number, the date of test, origin, net content, name and address of the labeler or seller, and a statement that the seed meets a standard for quality factors (see Figure 14).

It is also possible to label commercial seed by grades (Grade 1, Grade 2, etc.), with or without detailed information. A grade system alone is not as precise as detailed labeling in informing a purchaser what is being bought. It also does not reward the producer or seller with the highest quality product. A grade system of labeling can include one or more quality factors with or without additional detailed labeling. A grade system combined with detailed labeling would satisfy both a less educated purchaser and a sophisticated purchaser and would make education and extension easier.

Labeling that includes information regarding health standards or the purpose of the seed treatment might also be considered.

Sampling. Instructions for sampling by seed law enforcement technologists should be adopted from the International Rules for Seed Testing of the International Seed Testing Association so the instructions can be changed as the ISTA rules change.

Registration of seed sellers. Registration is one means of recognition and control of seed sellers. Beginning with a simple registration system without a fee will encourage the development of seed sellers and provide a way to determine where educational efforts and control must be directed. The status of the seed seller in a community may improve if he has a certificate of registration. As a less involved alternative, a seed law enforcement unit could compile an informal list of persons who customarily sell seed.

Figure 13. A Certified Seed label combining the information essential for seed certification with analysis details that might be required by a Seed Act

CERTIFIED SEED

Certified by: National Seed Certifying Authority

CLASS OF SEED: _____

KIND: _____

VARIETY: _____

LOT No : _____

GERMINATION (MIN.) _____ %   INERT MATTER (MAX.) ____ %

HARD SEED (MAX.) _____ %   OTHER CROPS SEED (MAX.) ____ %

PURITY (MIN.) _____ %   WEED SEED (MAX.) ____ %

DATE OF TEST: _____ TAG VALID FOR (YEAR): _____

POISON, TREATED WITH _____

(DO NOT USE FOR FOOD, FEED OR OIL PURPOSES)

CERTIFIED SEED PRODUCER CODE No. _____

NAME: _____

ADDRESS: _____

The seed to which this label is attached is equal to or above the minimum
requirements for this class of certified seed as specified in the Certified Seed
Regulations and for the requirements as given in the Seed Regulations.

NET CONTENT: _____ KGS.

Figure 14. A simplified analysis label. Used when minimum requirements are established for seed quality, but details are not required on the label. If a grade system exists, the grade designation would appear on the label.

ANALYSIS LABEL

NAME:_____

ADDRESS:_____

KIND:_____

VARIETY:_____

LOT No._____

DATE OF TEST:_____

ORIGIN:_____

The seed to which this label is attached is equal to or above the minimum Ministry of Agriculture requirements as specified in the Seed Regulations.

POISON, TREATED WITH_____

(DO NOT USE FOR FOOD, FEED OR OIL PURPOSES)

NET CONTENT: _____ KGS.

Registration of varieties. This may be a part of the system to control the quality of seed that is sold. When registration is mandatory it is usually part of a pre-marketing control system. Official registration of varieties is based on the idea that a government must protect the farmer even to the extent of determining which varieties can be made available for purchase. The owner or introducer who wishes to sell a variety must apply to the government, pay a fee, and submit a sample for official agronomic tests. The tests are then conducted at one or more locations for two to five years to determine whether the variety warrants being made available to consumers. This decision may be made on the basis that the variety is suitable for crop production or is at least equal to varieties already on the market. Alternatively, the growing tests of plant breeders may be officially observed instead of conducting separate official growing tests. Or the results of tests made by other official agencies might be accepted.

When private research is involved, legislation might require that seed of any new variety be made available to the government's variety evaluation unit so that tests can be made and the results published. However, the sale of the variety would not be delayed until such tests were completed (presumably the introducer of the variety would have tested it sufficiently to justify its introduction and use).

If registration is not a part of the system adopted, a list of recommended varieties or a list of varieties on the market, or both, may be developed. This system reduces a government's responsibility and simplifies procedures. Many governments find they have higher priority uses for their limited resources than expending them on extensive official testing for variety registration. Although a farmer's acceptance of new varieties rests upon many factors, his own evaluation of a variety is the ultimate criterion for acceptance. Therefore, the "official" assessment of varieties that originate from sources other than public-sector research needs to be handled as simply as possible.

Imports. Control of the quality of imported seed should parallel the control of domestic seed. If seed on the domestic market is controlled, the quality of imported seed normally should need no additional control.

False advertising. This aspect of seed control is significant to the degree that free enterprise exists. If seed is advertised for sale, seed legislation should include provision for penalizing false and misleading advertising. For example, advertising that claims "high" yields for varieties when those claims are unsupported by scientific trials is false and misleading.

Record keeping. The enforcement of seed laws requires documentary evidence to establish intent when falsely represented seed is delivered to a consumer. The "intent" could be a mistake or a willful violation. Thus seed legislation should require all seed sellers to keep detailed records, by lot, of seed purchases, sales, tests, labeling, and treatment, if any. The records should also include a representative sample from each lot handled.

Exemptions. Every seed law should have stated exemptions to ensure that products sold in the form of seed but intended for other purposes are not illegal. Wheat or maize for milling purposes is an example. Also, farmers who sell part of their own production to another farmer should be exempt, provided they do not transport the seed from their farms or advertise it. Organizations that transport seed and other commodities as a business should also be exempt. When a person sells seed that he has purchased, he should be exempt with respect to the correctness of the variety name if the variety is not distinguishable by appearance and is not as it was represented to him. (Proper records should be kept to indicate who should be held responsible.) Seed for experimental or research purposes may also be exempt.

Seed testing responsibilities. This section can establish official seed testing responsibilities, require that Certified Seed be officially tested before being marketed, specify who is responsible for testing samples as part of a marketing control program, and establish seed testing rules. Service testing for farmers, seed enterprises, and other seed sellers on a fee basis can be officially recognized. Commercial or privately financed laboratories may be encouraged to do the primary testing of commercial seed, but they need not be included in this section.

Tolerances. This section should establish the basis for recognizing that seed is not a uniform biological product and that two tests of the same sample seldom yield identical results. Tests made by two different people on two different samples from the same lot are even less likely to agree exactly. Tolerances can be recognized in the Seed Act and included in the rules and regulations by reference to the tolerances in the International Rules for Seed Testing.

Variety control plots. Planting seed from samples collected for seed law enforcement in variety control plots can be a useful supplement to seed tests, which usually cannot verify the accuracy of the labeling with respect to variety. Specific legislation for variety control plots may be unnecessary however.

Exports. It may be argued that if every country controls imports, no need exists for the control of exports. If exporting seed is economically significant to a country, however, it might choose to control the quality of the seed exported. This may be done by requiring official lot sampling, or testing of submitted samples, and by issuing International Seed Analysis Certificates in the case of members of ISTA or domestic seed analysis certificates.

Plant variety protection. The purpose of plant variety protection is to grant a plant breeder exclusive rights to market a variety that he has developed. Any other person who wishes to market the protected variety must be licensed by the developer or owner and pay a royalty. Usually plant variety protection is the last legislation to be considered. However, some countries have adopted such legislation at an early stage in order to encourage foreign investors to introduce new varieties and to stimulate private plant-breeding programs. Such legislation may be part of an all-encompassing seed law, but in most countries it is a separate law. Basic forms of plant variety protection legislation can be obtained from countries that have such laws or from the International Union for the Protection of New Varieties of Plants (see Appendix E).

Two basic procedures exist for determining whether a variety is new and novel (distinct, uniform, and stable) and entitled to exclusive rights. Countries that already have established a variety registration system with accompanying official evaluation tests may find it easier to include testing for novelty in the testing for variety registration. Countries that do not have a variety registration system or that do not conduct official tests for novelty have accepted the breeder's tests to verify the novelty of the variety. Some countries use both systems: in some crops or some circumstances official tests are made and in others the breeder's test results or results reported by other official stations are accepted. Under either system, the characteristics of the existing varieties must be established before determining whether a variety is new or novel. The magnitude of the task is so great it is wise at first to accept only a few crops for variety protection. The characteristics of existing varieties, obtained by growing-out tests or by a literature search, can be placed in a computer to simplify later searches for novelty.

Plant quarantine. Plant quarantine is in a separate category from other seed import regulations. This legislative requirement may be quite restrictive in order to bar certain alien diseases and insects. But plant quarantine controls should proceed step by step. They should not impede the movement of seed unless there is a sound scientific basis for doing so. It is excessive to forbid seed imports from areas where quarantined diseases and insects are not known to exist. Quarantine restrictions for diseases already prevalent domestically are a dubious practice. If a disease exists in the seed's country of origin but is not prevalent within the importing country, a practical control system is to (a) have the seed crop properly inspected in the field while it is growing in the originating country, (b) devise laboratory methods to detect the disease in the seed, or (c) plant the imported seed in isolation and inspect it. Priorities must be set, technicians properly trained, and facilities developed if a quarantine system is to be an asset rather than a liability. Establishing plant

The growth of quality control components

QUALITY CONTROL	Stage 1	Stage 2	Stage 3	Stage 4
Government administrative orders	**	*		
Seed certification		*	***	******
Seed testing		**	****	********
Legislation: marketing control				***
Legislation: seed testing			***	***
Legislation: seed certification			***	***
Legislation: plant variety protection				*
Legislation: plant quarantine		*	***	***

quarantine regulations on a regional basis should be considered. The European and Mediterranean Plant Protection Organization, for example, has made procedures for the classification of pathogens considerably more uniform among countries in its area.

Seed certification. Before seed legislation encompassing all facets of a seed industry is considered, seed certification may have previously been established under other legislation and be in full operation. If not, there is justification for legislation to control the seed certification system to ensure that certified seed is what it is represented to be.

General provisions. Each law may have provisions for delegation of duties, liability of public employees, authority for rules and regulations and public hearings, inspection of seed, education, seizure or stop-sale of seed, penalties, appeals, authorization for funds expended, cooperation with other domestic and international agencies, and the effective date of the legislation.

Possible wording for a General Seed Act and a Seed Certification Act is given in Appendix D. The International Union for the Protection of New Varieties of Plants can supply plant variety protection guidelines. Acts dealing with plant quarantine must be tailored so closely to the insect and disease conditions of individual nations that it is difficult to generalize about them.

ORGANIZATION OF QUALITY CONTROL PROGRAMS

The way quality control programs are organized can make the difference between success or failure. Starting a new program has its advantages because no poor precedents have been established. However, deciding how to organize seed certification, testing, and other quality control activities for a totally new program can be confusing.

Before considering the alternatives for organizing quality control programs the operational objectives should be stated. There should be no conflict of interest between production and quality control activities. Resources, including vehicles, should be adequate for flexible action. An adequate, competent, and responsible staff should be developed and maintained. Regulations should be applied with consistency and uniformity. A service-oriented attitude should prevail in applying the rules. An understanding of and dedication to good quality seed should be developed among seed enterprises, seed sellers, and farmers. Operations should be based on realistic seed standards. Fees from customers or funds from the government, or both together, should be sufficient to pay for an effective program. Educational activities should receive adequate emphasis. And seed quality activities should have international acceptbility.

Organizational Alternatives

Seed certification, enforcement of seed marketing laws, and seed testing are organized in various ways around the world. In a developing country, certification might start at a plant breeding station, but the needs will soon exceed the station's resources. Some countries have a central authority for the three functions, thus the units are closely integrated. Other countries have separate agencies for each function. Some countries even have separate organizations for different species.

When establishing a new seed program, especially a limited one, it may be most practical to place seed certification, enforcement of seed marketing laws, and seed testing under a central authority at a single location, such as a national seed center (Figure 15). This agency should have no production interests and should be as autonomous as possible. Provision could be made in the legislation for the minister of agriculture to delegate quality control responsibilities to such an agency. Within the agency, as much responsibility as possible should be delegated to

Figure 15. Seed quality control for a modest program organized under central authority

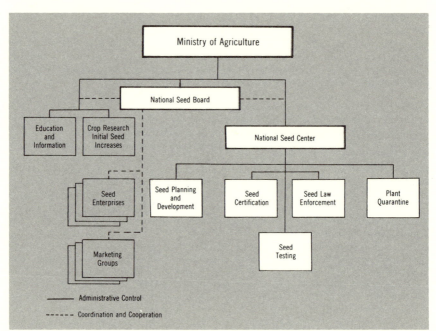

facilitate decision making and to avoid bureaucracy.

A plant variety protection (the plant breeders' rights) unit, if organized, and a plant quarantine unit could be located at the national seed center. But in large countries or for large programs, state or provincial seed centers with a national coordinating mechanism may be appropriate. Excessive central management of decentralized activities should be avoided.

The number of seed testing units should be kept to a minimum in order to make seed testing as economical as possible and to reduce variability in results.

As an alternative to a single agency for certification, enforcement, and testing, an autonomous authority could be created for seed certification alone (Figure 16). (The proposed Seed Certification Act in Appendix D includes the possibility of the minister of agriculture delegating this authority to a separate agency.) Under this arrangement, the enforcement of seed marketing laws could become a part of a larger quality control program for fertilizers, feed, pesticides, and other agricultural products. Seed law enforcement has much in common with other mandatory

Figure 16. Seed quality control for a larger, more advanced program organized to give seed certification greater autonomy

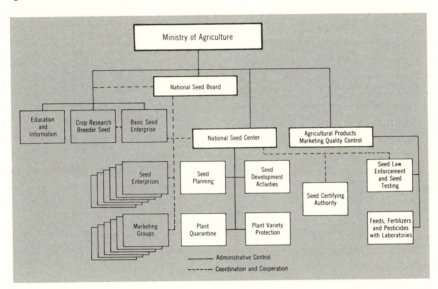

quality control activities and, therefore, could be carried out as part of such a group in the ministry of agriculture. The physical location of the office for seed marketing law enforcement could nevertheless be at the national seed center along with seed certification and seed testing. In larger programs, especially, it is advantageous to have seed certification organizationally separate from enforcement of seed marketing laws. Separation reduces confusion among seed growers and seed enterprises and averts possible staff conflicts in achieving the objectives of seed certification and mandatory seed law enforcement. Seed certification as a service can become financially self-sufficient whereas seed law enforcement for marketing control is a protection for consumers and usually receives more public financial support. Thus, the financial management of the two activities may be quite different.

National Seed Board

A national seed board should be responsible for the total seed program including the national seed center (see Chapter 1). This board could recommend policies and develop proposals for financing, locating, and operating the certification, law enforcement, and testing units. Technical subcommittees that work on these questions should include the leaders of the national seed center.

If a separate seed certification agency or authority is created, the national seed board could form a special committee to formulate seed certification policies and approve standards.

MANAGEMENT IN QUALITY CONTROL PROGRAMS

Weak management can destroy even well-organized quality control programs. To guide administrators, managers, and heads of seed certification, testing, and law enforcement programs, several specific activities are listed below. Sections on personnel and their educational requirements are included to help decision makers build a competent staff.

Managing Seed Certification

The seed certifying unit or agency inspects seed crops in the field, samples processed seed lots, assesses testing results relevant to individual seed lots, and authorizes labeling containers of seed as certified. A fundamental responsibility of the head of seed certification is to develop a competent staff. Individuals must be selected who are highly motivated, capable of working effectively with seed growers and seed enterprises, prepared to travel to accomplish their duties, and aware of the importance of thorough, objective work. The staff should receive adequate training and educational materials so they can identify all varieties being certified, the diseases included in the standards, and weeds of importance in seed. In addition, the inspection staff must be familiar with harvesting, drying, processing, and storage techniques so they can advise seed enterprises. (If "approved seed processors" are to be recognized, specific instructions on inspecting their operations might be prepared for the staff.)

Any good manager closely supervises the work of the staff. The head of seed certification should assign technologists to specific areas of inspection and ensure that they arrive sufficiently ahead of critical inspection periods so the work can be completed on time. The manager should visit the seed certification technologists regularly to uncover problems, guide the inspection staff, and be aware of the details of the work being done. The inspection staff should advise the manager of fields likely to be rejected so he can visit those fields if necessary.

Policies and procedures must be established. The head must see that seed certification standards are achievable and published and that procedures for field and seed inspections are clear to the staff. Outlines must be prepared on how seed growers and seed enterprises should apply for seed certification, and they should be supplied with guidelines and advice on meeting the standards (see Appendix C). A procedure for officially rejecting fields must be developed, perhaps with the assistance of the seed certification committee if serious problems develop. A system for maintaining records on all seed lots certified must be created. (The records are used to continue the identity of seed lots of each variety from one generation to another, to verify the varietal purity of the source of seed used, and to designate the generation—that is, whether the seed in question is the first, second, or third generation from the Basic Seed.)

Problems with supplies and equipment that might hamper staff work must be avoided through careful planning. The staff needs materials and instructions for seed sampling to ensure timely seed testing and approval or rejection of seed lots. Forms for recording the results of seed inspections (see Appendix C) must be on hand when needed. Certification labels, and perhaps seals, must be obtained far enough ahead of time so the staff can provide them for seed lots that have passed certification. And the staff must have some means of transportation to inspection sites.

Finally, the head must ensure that the seed certification unit or agency is adequately financed and that funds are available when needed. Arrangements must be made to investigate complaints from users of Certified Seed and to cooperate fully with the seed law enforcement unit on violations of the Seed Act. The head is responsible for building good relations with others and must foster close contact with crop research, Basic Seed multiplication, seed testing, seed law enforcement, and extension units as well as with high-level government administrators, universities, seed growers, seed enterprises, and seedsmen's associations. And the head must maintain links with neighboring and international certification groups.

Managing Seed Testing

The head of a seed testing unit should be well acquainted with testing procedures. Also, the unit needs specialists knowledgeable in purity analysis, germination, moisture determination, and, possibly, seed health, seedling vigor, and varietal purity testing. In addition to these specialized skills the staff should be familiar with taxonomy, seed processing, seed treatment, and seed storage.

Since seed testing is often seasonal with one or two periods of peak work load, the size of the staff must be carefully considered. The number of workers should be related to the species involved and to the efficiency and experience of the staff. Seed of crops like maize, beans, and wheat takes less time to analyze than tiny grass seed. It is also desirable to have some seed testing technologists in training. When deciding staff size, the time required for nonroutine activities such as practical research and education must be considered. Other work such as varietal plot testing needs to be considered for low work-load periods. According to studies in Europe, eight trained analysts plus office help can do purity, germination, seed health, and moisture tests on five thousand samples a year. The budget must allow for the employment of some part-time workers during peak periods—a sample should be ready for a germination test within twenty-four hours after it reaches the laboratory. If varietal plot testing is organized, staff members must be experienced in growing the crop.

The head of a seed testing unit is responsible for properly planning and supervising all activities to maintain the accuracy and uniformity of tests. For example, the head of seed testing needs to be sure that bags for collecting seed samples are printed and distributed in advance to facilitate obtaining certification, service, or seed law enforcement samples and to facilitate the recording of sample information at the seed testing laboratory. A supply of cards or a record book is needed to systematically maintain information on samples received (see Appendix C). The head of the laboratory will probably find it useful to organize the records for official, certification, and service tests separately and to use color codes to identify each group. Cards on which analysts can record test results must be obtained. The cards can be filed for answering questions about the seed tests made. If the head of the laboratory is concerned about impartial test results, the use of numbers on these cards instead of names can reduce possible bias.

The head of the laboratory must see that equipment is adequate and well maintained and that work is soundly organized. For example, systematic procedures include properly dividing a sample to obtain working samples and promptly routing those samples to the different sections for the tests requested or required. The head needs to ensure that purity testing is done by a person knowledgeable about crop and weed seed characters and that the laboratory has a good seed collection for use in making comparisons in case of doubt. He needs to ensure that the equipment for germination testing is operating properly and that correct temperatures are being used. The head must give sufficient guidance to the staff so that evaluations of seedlings are consistent, for example, by providing copies of publications such as the *ISTA Handbook for Seedling Evaluation*. (Also, the ISTA International Rules for Seed Testing list the common abnormalities in different species.) Determination needs to be made whether special seed health tests are required and whether the facilities and pathologists can be obtained to conduct such tests. (The Plant Disease Committee of ISTA offers training in laboratories of member countries and through special sessions. The Danish Institute of Seed Pathology for Developing Countries and some seed testing stations offer special training. ISTA has several publications on testing methods and the identification of seed-borne diseases to help standardize and facilitate such training.)

A critical responsibility of the head of the laboratory is to see that test results are calculated accurately, checked, and entered on the special working cards and that each test average is compared with the individual results to determine if the variation exceeds tolerance levels. If the results are outside the accepted tolerances, he needs to be certain that a retest is made. (The individual test results may deviate from the calculated average—a certain variation is always present due to the individual characteristics of the seed represented in the different test replicates. A tolerance factor is needed in order to avoid unnecessary retesting.) Ultimately, the laboratory head must ensure that accurate, neat reports of the results are sent to those who supply the samples (see Appendix C for a laboratory report form).

In addition to the regular seed testing facilities, the head of the laboratory must make sure that samples that have been tested are stored properly so that, if necessary, they can be retested to answer questions about the results.

Time is important to the customer. The time required depends on the test to be made and the kind of seed, but the head of the laboratory must monitor the time from the receipt of samples to the sending of reports and help the staff become time conscious. In addition to providing leadership, the head of the laboratory meets those using the laboratory's services and, therefore, must be a good public relations officer.

Managing Seed Law Enforcement

Consumer protection is the primary purpose of seed marketing laws. The level of protection possible is related to funding. A seed law enforcement unit usually depends upon a central source for financial, personnel, and legal services. The head of the unit must plan the program to get the most good from the funds and skills available. Competent, trustworthy seed law enforcement technologists (inspectors) must be hired to sample seed for sale. The number needed depends upon the amount of seed to be inspected, the area to be covered, and the availability of transportation for staff members.

Seed inspection is often only a seasonal occupation. Especially in small new programs, combining seasonal seed inspection for seed law enforcement with seasonal certification field inspection, control plots, or seed testing may make it possible to keep well-qualified people occupied year-round. Or, in some situations, it may be economical to use the same personnel to inspect several commodities being sold in the same locality.

The head of a program must be sure that the staff has transportation to visit seed enterprises and seed sellers. Periodically, the manager should travel with the law enforcement technologists to keep abreast of their problems and to offer guidance.

The head must be sure that staff members are knowledgeable about seed quality and the Seed Act, since in the beginning of the program they may do more educational work than inspection work. In-service training and regular workshops should be organized to increase the staff's knowledge and skills, particularly if inspectors are not working out of a central location. He must ensure that the staff is equipped to properly draw a sample to send to him or directly to the seed testing laboratory together with details about the sample. (If the staff is sufficiently qualified, the head can permit them to make a preliminary purity analysis or weed identification on the spot to guide seed sellers.) Seed inspection report forms (see Appendix C) must be obtained and provided. Examples of seed inspection report forms can be obtained from countries already inspecting seed in the market (see also USDA's *Instruction 961-3*, AASCO's *Seed Administrator's Handbook*, and *A Handbook for Seed Inspectors* published by the National Seeds Corporation of India).

The leader of a marketing control program should devise a clear stop-sale procedure for law enforcement technologists to use to prevent the sale of falsely labeled seed or seed that is unfit for planting until the labeling is correct or the seed is disposed of according to law. A stop-sale form should be provided (see Appendix C). The steps needed when seed should be removed from the market because it is unfit for planting should be indicated, such as seizure by court order and disposal by grinding for feed, or burning if necessary (to destroy harmful weed seed that cannot be removed by cleaning). The head should develop a procedure for publicizing the names of those who violate the law since this is one of the best deterrents for habitual violations. In summary, the unit's head should develop a series of steps for use in situations involving violations as illustrated in Table 3.

The head of a law enforcement program must give the staff guidelines for developing a listing of seed enterprises and seed sellers, or for registering them; suggestions on the number and kinds of educational meetings to be conducted; and suggestions on the educational procedures to use with seedsmen to obtain compliance with the Seed Act. Finally,

Table 3.
Seed law enforcement procedures

Kind of Violation	Example	Suggested Action by Seed Law Enforcement Technologists
Technical	Seed quality is not affected but labeling is incorrect	Labels can be corrected as specified by the seed law enforcement technologist
Minor	Differences between components and information on the label are just beyond ac-acceptable tolerances	A stop-sale procedure can be used, label corrected or seed recleaned, and a warning letter issued
Serious	Violations that will do serious harm to the grower such as extremely low germinating seed or seed containing prohibited weed seed	Violator can be given an opportunity to explain violation to the head of the seed law enforcement unit; violator may be taken to court; seed might be seized by court order and removed from the market

the head must be sure the staff has instructions for taking samples, completing reports, and submitting samples to the seed testing laboratory, and for following up on complaints received from seed users.

Because varieties of seed can seldom be distinguished by seed characters, the head of seed law enforcement may wish to consider selecting a random group of official samples from the seed testing laboratory to plant in the field the following season to see whether seed of selected species are being correctly labeled as to variety. He may enlist the help of the head of seed certification or of the seed testing laboratory depending upon staff availability, staff qualifications, and facilities.

Personnel

Adequate technical facilities and a qualified staff with differing levels of education are necessary to start a seed quality program (see Chapter 7). All personnel must understand the aims of a seed program and the justification for those aims. The categories of personnel involved and the information they require are:

High-level officials: policies, budget and legal aspects, and some background on major sectors of the seed program
Certification, testing, and law enforcement management: seed grow-

ing techniques, crop and varietal morphology, rules and regulations, sampling, labeling, seed testing, extension, administration

Seed certification technologists: seed growing techniques, crop and
varietal morphology, certification rules, field inspection techniques, sampling rules, seed testing principles, labeling, information techniques, administration principles

Seed testing technologists: seed testing techniques, rules and regula-
ations, sampling, labeling, information techniques, administration

Seed law enforcement technologists: seed legislation, sampling,
labeling, testing, certification, information techniques, administration principles

Seed growers: seed growing techniques, certification rules, seed
quality understanding (see Chapter 4)

Seed enterprise personnel: seed growing principles, certification
rules, seed quality understanding, processing, marketing, distribution, transportation, management (see Chapters 4, 6, and 8)

Educational Requirements

The education and in-service experience necessary for the different
levels of personnel in certification, law enforcement, and testing are as
follows:

The heads and the chief officers of the units need academic training in
seed technology, agronomy, or plant breeding with courses in botany,
crop production, plant breeding, plant physiology, plant pathology, and
information techniques. Each head and his deputy also should have some
administrative and practical experience. It is an advantage if they also
have experience in other facets of the seed program.

The seed certification, testing, and law enforcement technologists need
backgrounds in agronomy, botany, and plant breeding plus broad
knowledge about the seed program. The amount of knowledge required
depends upon the level of employment. The main emphasis of in-service
training should be on the duties to be performed. It may take trainees
considerable time to gain the competence needed. The best stimuli for
such personnel are clearly defined responsibilities and praise for work
done well.

The lower-level personnel in the seed testing unit need extensive training before they can make reliable tests. It may take as long as three years
to learn all the skills associated with the work.

Seed Quality Control: Major Policy Points

1. Level of emphasis that is to be placed on seed quality both inside and outside government activities
2. Broad approach for checking quality of local and imported seed that is being distributed
3. Whether seed certification is to be initiated, the manner of organization, and how a practical level of standards is to be established
4. Kind of seed quality legislation needed, the details to be included, when it should be enacted, and how it is to be enforced
5. Number, size, and location of seed testing laboratories needed
6. Organizational structure and interrelationships needed for all quality control activities supported by government

---

REFERENCES

Association of American Seed Control Officials. 1976. *The Seed Administrator's Handbook*. Richmond, Va.

Association of Official Seed Certifying Agencies. 1971. *AOSCA Certification Handbook*. Clemson, S.C.

Chalam, G. V.; Singh, Amir; and Douglas, J. E. 1967. *Seed Testing Manual*. New Delhi: Indian Council of Agricultural Research and USAID.

Feistritzer, W. P., ed. 1975. *Cereal Seed Technology*. Rome: FAO.

International Seed Testing Association. 1976. International Rules for Seed Testing. *Seed Science and Technology* 4:1–180.

Jenkins, M. T. 1953. Problems Facing the Expanded Use of Hybrid Maize in Europe and the Middle East. *Sixth FAO Hybrid Maize Meeting Report*. Rome: FAO.

National Seeds Corporation. 1972. *A Handbook for Seed Inspectors*. New Delhi.

U.S. Department of Agriculture, Grain and Seed Division. 1976. *USDA Instruction 961-3*. Washington, D.C.

Wellington, P. S. 1969. *Handbook for Seedling Evaluation*. Zurich: International Seed Testing Association.

6
Getting Seed of
Improved Varieties Used

No benefits result from making good quality seed available unless farmers obtain and plant seed of improved varieties. Too often seed programs focus on producing and processing seed and neglect factors that contribute to seed use. Administrators of both public and private agencies must be sensitive to factors that influence farmers' acceptance of new varieties. They should establish mechanisms to broaden the public's knowledge about seed and improved varieties, and they should encourage a sound marketing system to develop. Finally, government administrators need to develop clear policies for supporting activities that increase the use of good seed.

INFLUENCES ON ACCEPTANCE OF IMPROVED VARIETIES

Although producing and introducing good seed of new varieties are largely technical and economic issues, the adoption of improved seed, like any new farming practice, is directly linked with what farmers know, what they understand, how they feel, and what they are willing and able to do. Contrary to common belief, farmers as a group do not resist change. In fact, they have survived over the years because they have adapted to changes in their physical, economic, and social environments. In recent years, farmers in many parts of the world quickly adopted new high-yielding rice and wheat varieties and maize hybrids when it was clearly advantageous and physically, financially, and socially possible to do so.

According to John Gerhart, when hybrid maize was introduced into western Kenya in the 1960s the rate of adoption by both large and small farmers was faster than the rate of adoption by farmers in Iowa in the United States when hybrid maize was introduced there. The first small-scale Kenyan farmers to adopt hybrids did so against the advice of the extension service. The extension service was recommending synthetic varieties, fearing that farmers would not pay the price for new hybrid seed every year and would plant F_2 generation seed. In this case, at least, the farmers were less conservative than the officials paid to introduce change.

Seed of a new variety is rarely adopted without some change in cultural practices. Perceptions about the cost, effort, or risk associated with the new practices often block ready acceptance of a new variety. In the Philippines in 1967, initial introductions of IR8 (the first variety named by the International Rice Research Institute) were hampered by critics who tended to confuse the variety's ability to respond to high levels of fertilization with the notion that the new rice *required* large amounts of fertilizer as well as farming skills beyond the capabilities of the small rice farmer. This misconception persisted until the farmers learned through experience that the new variety generally would do as well as traditional varieties without more fertilizer or unusual attention and that the yield increase when more fertilizer was used would result in levels of profit such as they never before had known.

This example illustrates the importance, in planning a campaign to introduce new varieties, of identifying the audiences to be informed. It is not sufficient to inform just the farmer; others not similarly informed are likely to be negative, particularly if they have faulty ideas about the changes that might result. Among those that should be kept advised of the availability and potential of new varieties are landowners, lenders of money, sellers of farm supplies, buyers and processors of the crop, operators of transportation and storage facilities, and the general public, especially if the crop is consumed locally.

A campaign to introduce seed of a new variety must be planned to last three to five years. Although a variety may be "big news" when it is announced, adoption of the seed generally spreads over several years. Moreover, producing and distributing enough seed in one year to achieve a near total adoption throughout a country would be difficult.

Finally, a farmer's acceptance or rejection of an innovation results from his perceptions of it rather than from its physical and economic realities. Until the farmer has seen others use the innovation, or tried the innovation himself, his perceptions (and their roots) must be taken into account in order to establish effective communication. Change or modification of those perceptions may be the first and possibly the only communication task.

Factors Farmers Consider

Social scientists who have studied why farmers adopt new varieties and agricultural practices have delineated the characteristics of an innovation, such as seed of a new variety, that determine whether and how rapidly a farmer will accept it.

Relative advantage is the degree to which a farmer perceives that the improved seed or new variety will raise benefits, or lower costs, as compared with the benefits, or costs, associated with the seed or variety he currently uses. Although "relative advantage" usually is interpreted as profitability, it may take the form of a difference in effort, risk, prestige, or social approval.

Reliability is the degree to which a farmer perceives that use of the new variety will consistently produce the minimum crop needed to feed his family and to provide the income he normally requires from sales.

Simplicity is the degree to which a farmer perceives that seed of a new variety and its associated practices are easy to use.

Compatibility is the degree to which a farmer perceives that a new variety is consistent with his needs, values, past experience, and farming situation. A variety with an extremely long or short growth cycle, for instance, might not be compatible with a farmer's cropping system or a community's cropping pattern. It might cause serious problems in the use of irrigation, control of insects and birds, access to labor for cultivation and harvest, or availability of equipment for marketing, processing, or transportation.

Visibility is the degree to which a farmer can see the results of having used a new seed, and how apparent they are to others. Better germination may be one of the first things a farmer notices about improved seed. If the new variety and old variety have distinctly different growth characteristics, early visibility will be possible; but if they do not, differences must be apparent in the quality and volume of the harvest.

Divisibility is the degree to which a farmer perceives that he can try an innovation on a limited basis. Seed has a distinct advantage over some other innovations in that a farmer usually can limit his trial planting to a small fraction of his land.

Independence is the degree to which a farmer perceives that he can decide to adopt an innovation without consulting anyone else. Unless landlords, credit institutions, or the community impose demands or restrictions, the decision to use seed of a new variety can be made independently. If the seed is readily available, farmers generally are free to elect to plant it. Many other innovations may have a lower degree of independence. A farmer may not be able to obtain irrigation unless enough other farmers agree to help develop and use a system, or the use of fertilizers or other inputs may not be profitable unless seed of a high yielding variety is available.

Importance of Suppliers and Markets

To be able to use improved seed, a farmer must find it available at a fair price, at an appropriate time, and at a convenient place in the quantities needed and in manageable units. But, in addition, he must have access to supplies such as fertilizer, pesticides, and equipment as well as the money or credit to pay for them. A farmer may also need help with transportation and storage of supplies, and with handling the crop at harvesttime and during marketing.

These factors are particularly important in programs for the small farmer who is less likely to have experienced the wide range of decisions involved in successfully adopting new production technology than is a larger producer. The small farmer receives less attention from extension workers and production specialists and, at the end of the season, is more at the mercy of the marketing system. Consequently, it is important to take into account the characteristics of farmers.

Characteristics of Farmers

When campaigns to introduce new varieties or other technology to farmers seem, after a few years, to be less successful than anticipated, the leaders of a campaign usually blame the resistance of farmers. They rarely question the suitability of the technology for the farmers or the methods employed in introducing it. Concern for the farmer must start earlier. When designing and testing a new variety, efforts must be made to analyze the characteristics of the particular groups of farmers to whom the seed will be introduced. Audience analysis, in the sense used here, involves examining the farmers' knowledge, understanding, acceptance of sources of information, attitudes and beliefs, perceptions of effort and reward, and ability to act independently.

Knowledge: how much a farmer knows about varieties and differences in seed quality, about the quality of his own seed, about identifying good seed being sold, and about the sources of seed of new varieties.

Understanding: how well a farmer understands the various seed terms used, how a new variety differs from the variety he has been using, and the importance of associated cultural practices.

Acceptance: what sources of information a farmer regards as credible; what sources of seed he regards as reliable and trustworthy; whose opinion he usually seeks about new seed and other farming practices; how much credibility government workers, representatives of commercial firms, and neighboring farmers have.

Attitudes and beliefs: a farmer's ideas, notions, customs, and rites relating to nature, farming, and seeds. (For example, one production specialist tells of a group of farmers who only wanted "a little bit" of seed of a new rice variety. When he asked why, he learned they wished to mix it in with their usual rice seed, believing that they could, over time, upgrade their rice in the same way they had been taught to introduce a purebred rooster into their flocks to improve their chickens.)

Perceptions: a farmer's perceptions of the effort involved in using a new variety in relation to the rewards expected. "Effort" may be defined as the cost, labor, or risk involved, or the possibility of criticism or ridicule by neighbors. Rewards may be monetary gain, reductions in labor or risk, or gain in social prestige. It may be possible to help farmers arrive at more realistic expectations, or at least to reduce the perceived amount of effort in relation to the expected reward. Similarly, it is useful to recognize that while most efforts are immediate, most rewards are delayed; when it is possible to postpone some effort and make some rewards more immediate, farmers become more receptive.

Ability to act: farmers may be unable to adopt new varieties or technology for reasons beyond their control. They may not learn about the seed of a new variety in time for current plantings. Some who know about the seed may not know where to get it, or may lack the skill to handle associated practices. But beyond this, some farmers may not be able to act because of physical, financial, or social constraints, many of which might have been alleviated through communication, conversation, and demonstrations.

Variables Influencing Rate of Adoption

Those planning to produce and introduce seed of a new variety must estimate how rapidly farmers are likely to adopt the innovation. If planners are too conservative, inadequate distribution will create frustration among farmers who are unable to get enough seed. On the other hand, it is costly to produce more seed than farmers are willing to adopt initially or than the distribution system can handle.

The rate of adoption—normally defined as the number of farmers who adopt a new idea in a specified time—is influenced by the perceived attributes of an innovation. One analysis of hundreds of studies of innovations involving thousands of farmers around the world indicated that perceived attributes explained more than half of the variance in the rate of adoption.

A second influence on the rate of adoption is the type of decision involved. Generally, adoption is most rapid when a single authority can decide that a group of farmers will use a new variety. Adoption is less rapid when individual farmers must each decide. Adoption is relatively slow when collective decisions are involved, that is, when a majority of the social system's members must agree to act.

Third, communication channels affect rate of adoption, but in complex ways. Obviously, if no mass media exist, knowledge about seed of new varieties must pass from person to person, a slow process. But with complex innovations, or with complicated practices associated with the use of seed of new varieties, adoption may be promoted as much by farmers' conversations as by mass media.

The nature of the social system is another influence that should be considered in planning an information campaign for a new variety that is being introduced. In a social system in which other changes have been successfully introduced, adoption of seed of a new variety is likely to be more rapid than in systems where resistance to change must be overcome or unfavorable attitudes modified. It will be useful to identify important characteristics of the variety that can be emphasized with different groups.

Finally, the amount and nature of activity by the organization introducing the seed will affect the rate of adoption. Here again, success depends in part on the knowledge of the audience: some people need to be informed, others need to be persuaded, and many require instruction. When maize hybrids were introduced in Kenya, every packet of seed contained an advisory leaflet printed in English and Swahili. A farmer who could not read the leaflet frequently took it to an instructor. That was how many agricultural instructors and extension workers first heard about such recommendations as fertilizer rates, methods of application, and planting rates.

SPREADING KNOWLEDGE ABOUT IMPROVED VARIETIES

Several groups must work hand in hand to educate and inform farmers about seed of improved varieties. One is the crop research program that developed or identified the variety and tested it under various conditions, and thus has information valuable to farmers. Another is the extension program, which has crop production specialists, information specialists, or field agronomists skilled in communicating with farmers. (The extension program consists of educational and informational activities that may be organized into an extension service or a similar agency to help the farmers. The technical personnel may be called crop production specialists, farm-level workers, or something similar. Specialists in communication are usually part of such a group and may be called in-

---

Steps in Campaigns to Introduce Seed of a New Variety

Establish objectives
Test materials
Utilize local leaders
Offer direct experiences
Employ multiple communication channels

---

formation specialists.) A third group comprises the seed enterprises and marketing organizations that are interested in a new variety and may have staff members who can help transfer technology and information to farmers.

Wide differences exist from nation to nation in the ways information is spread about good quality seed and good varieties. Some of the systems to organize research programs and to bring research results to farmers are discussed in *To Feed This World: The Challenge and the Strategy* (by Wortman and Cummings). Seed is but one element in the complex of factors that must be considered when communicating with farmers. Thus, spreading knowledge about seed must be an integral part of the total system.

The most effective systems have found ways to focus research and extension resources on specific crop production objectives, to mount a national or regional campaign aimed at achieving those objectives, to have seed supplies available from several locations, and to sustain effort and leadership for enough years to have a noticeable impact on production. In some instances the crop research program has included a strong farm testing program that has been used not only to test varieties and technology but also to train farmers and extension specialists. In others, extension services have played a dominant role in conducting farm trials to spread knowledge and gather information to complement research findings.

The flow of information between specialists responsible for the development of a variety and those concerned with communicating information to the farmers is critical to the successful introduction of a new variety. For this reason some programs have crop production specialists or disciplinary specialists at the national or regional level who link these two groups.

To help formulate policy, to develop goals, and to coordinate crop production campaigns, a high-level board, council, or coordinating committee should be created. It should include the head of crop research and development; the head of the extension program; representatives from the national seed board, including at least one seed enterprise representative; and representatives of other agricultural supply organizations, credit agencies, and farmers. A national coordinator for each crop production campaign could play a valuable administrative role in implementing campaigns organized under a national crop production board.

The national seed board discussed in Chapter 5 could fulfill this role for seed only, but campaigns to introduce seed of a new variety with other accompanying supplies and technical guidance normally need a level of support that goes beyond the representation proposed for that group.

Establishing Program Goals

Governments encourage the use of seed of improved varieties to increase production and improve diets, to reduce imports of food, to earn foreign exchange through agricultural exports, or to raise the incomes of farmers. It is important to identify these ultimate socioeconomic goals and to be able to translate them into the specific intermediate agricultural goals on which realization of the ultimate goals will depend. In other words, the national crop production board should state what yield increases will be achieved in what amount of time, on how much land, and by what farmers. When specific goals have been established, activities related to seed production, information, and marketing can be synchronized to achieve the goals. Those responsible for informing and advising farmers and others will have a basis for determining what activities will be required. And a national coordinator or other administrators will be able to evaluate results, to see where the program succeeded and failed, and to identify persons and institutions responsible for the success or failure. But rather than seeking scapegoats, administrators should commend outstanding work and generate peer pressure on those who are holding back the program.

Communication with Many Audiences

Promoting the adoption of new varieties involves informing, persuading, and teaching more people than just the farmers. The activities of the groups that make up the total production system must be analyzed. This is particularly important if a campaign must change what various groups know, understand, accept, do, or are able to do. The introduction campaign must then identify the groups with which communication must be established and, for each group, what the principal communication objectives must be.

Throughout this process, the national coordinator needs to be sure that the communication specialists use their training and experience in support of the campaign objectives. The specialists can be called upon to make analyses and draft campaigns to create awareness, modify attitudes, obtain commitment, and stimulate action. Moreover, their help is needed, along with that of other specialists, to identify and mobilize all components of the seed program that must function to get seed produced, distributed, and used.

The national coordinator may wish to check the campaign plans to see how community resources are to be mobilized, how teaching and counseling will be provided, and what methods of evaluation are planned. Without evaluation, the campaign organization has little basis for knowing how well it is doing or where to introduce changes in future programs.

Principles of Effective Communication

Several communication principles are common to most successful crop production campaigns.

Establish objectives. For each audience, it is important to state specific behavioral or performance objectives. In other words, as a result of the campaign, how are the members of the audience expected to be different? What will they be able to do, at what level of proficiency, under what conditions? Unless objectives are clear and agreed upon by all responsible for achieving them, the chances of program success are low.

Test materials. Communication materials should be tested before they are printed in large numbers or broadcast widely. While it is not difficult to conduct such tests, they frequently are overlooked in the enthusiasm to get on with a program.

Find local leaders. It is important to develop rapport with community leaders because they can help organize programs, test materials, and act as channels for getting information to others. One local leader may not serve for all people or for all matters; leaders within various strata or cliques must be sought as well.

Give direct experience. Insofar as possible, communication activities should go beyond printed materials and broadcasts. To increase interest and reduce misunderstanding, the audience should have opportunities to see the variety and associated practices firsthand within the community, on the farm, in demonstrations, and in exhibits.

Use multiple channels. Appropriate channels of communication should be chosen. For instance, it is dangerous to give detailed instructions by radio. In print, a farmer can study the instructions in simple words and, if possible, in photographs or drawings. But radio can be effective in creating an awareness of the potential of new seed as well as how and where to get it. Radio is low in cost and readily available to farmers, and short messages can be repeated frequently.

Linking Research and Practice

Production goals will be attained more rapidly if research workers, extension and crop production specialists, seed suppliers, and farmers communicate and cooperate with one another. In traditional organizationaι structures where researchers, farm-level workers, and seed personnel work in different agencies, shared goals can help them overcome barriers. In some developing countries the process of getting good seed to farmers has been accelerated by encouraging farmers to participate in the research process, by stimulating the farmers' initial demand for seed, and by creating widespread public interest in agricultural productivity.

Encouraging Farmer Participation

Working with authorities of national governments, several international agricultural research centers have found ways to bring farmers into the agricultural research process. At the Centro Internacional de Agricultura Tropical in Colombia, for example, a panel of three hundred farmers is established for each crop to identify production problems of farmers. For twelve to eighteen months, trained interviewers regularly visit each farmer, note the production problems, and query the farmer. This methodology combines economic analysis with biological experimentation and generates information on the kinds of technology needed and most likely to be accepted, particularly by small farmers.

Another technique, which has been used at the International Rice Research Institute, is the "mini-kit." In the late 1960s, with the cooperation of government and industry, IRRI distributed packages to farmers that contained small lots of seed of up to ten experimental strains and established varieties. The recipients were asked to report their experiences and their preferences for the new strains.

Distribution of mini-kits involves farmers in the research process, makes them aware of new technology, and provides scientists with useful information about field performance in varied environments. It is unwise, however, to delay the release of new varieties until all returns on the performance of the experimental strains in the mini-kit have been obtained. Release of new varieties is best based on data acquired by the scientific staff in well-supervised trials plus early mini-kit results.

The mini-kit approach was introduced in Sri Lanka in 1970. Kits containing new varieties and selections that had demonstrated superiority in national variety trials were furnished free to farmers through extension channels. Farmers indicated their acceptance of a new variety BG 11-11, and subsequently one hundred thousand production kits containing a half kilogram of BG 11-11 seed, along with a set of instructions on how to grow it to obtain maximum yields, were produced and sold to farmers. Farmers who were pleased with the harvest were easily able to save some seed for their next planting. While BG 11-11 swept the rice-producing areas of the country in succeeding seasons, there was less demand than expected for Certified Seed of BG 11-11, probably because seed of the variety spread from the kits.

In locations where the seed production and marketing programs are not well established, production kits can serve as a short-term substitute for a developed commercial seed industry for some crops such as rice. However, the use of kits cannot replace the long-term need for an effective commercial seed industry distributing many kinds of seed throughout a country. Even in the case of a self-pollinated crop like rice, the seed of a new variety gets mixed and diluted with seed of other varieties when farmers save their own seed, so new supplies of seed are required. The development of seed enterprises capable of continually supplying good quality seed must remain as a long-term goal.

In Guatemala, the Instituto de Ciencia y Technología Agrícolas carries out the bulk of its research on varieties and practices in farmers' fields and uses experiment stations only for work on special problems.

Adaptive research is now used by many international and national institutions to validate new technology in differing environments. Generally, through organizations that operate at the local level, arrangements are made to place simple, replicated trials of new technology in farmers' fields in many ecological situations. Through such trials, scientists may discover, for example, the susceptibility of a promising strain to a disease or insect that was not present where the strain was developed.

All of these techniques permit farmers to participate in developing new technology. Furthermore, since the extension service is the most convenient channel for carrying out such projects, the extension worker learns how to manage the new technology. Nevertheless, well-organized demonstrations are needed to promote the acceptance of new seed and associated cultural practices. Trials with farmer participation provide the basis for designing demonstration plots that will succeed.

Thus, conducting trials in farmers' fields is a way to test new technology under a wide range of soil and management conditions and to synthesize information (from the farmers, from trials in their fields, and from experiment stations) into recommended practices.

Stimulating Farmer Interest

Although field trials involve some farmers, stimulating the interest of large numbers of farmers requires other techniques. Once the local suitability of a new variety or practice has been established through applied research, demonstration plots probably are the best tool available to local extension workers. The plots need not be large; it is better to have several small, well-distributed plots than one or two large ones. Nor is there any need for a research design with replications and controls. Generally, the varieties growing in surrounding fields provide the most appropriate comparison.

Local farmers can be consulted about where to place demonstration plots, taking into account the cooperativeness of the farmer who tills the land, access to the plot for cultivation and display, and the farmer's credibility with his neighbors. Demonstrations ought to be associated with any local programs of seed distribution or farmer training that may exist. In some communities more than one plot may have to be established in order to interest different or rival groups of farmers.

Farmers who receive mini-kits and participate in demonstrations may be the best candidates to be future seed growers and those around whom seed enterprises and marketing groups are formed. The field agronomist concerned with mini-kits and demonstrations should be closely linked to the seed technologist working for the development of seed production and marketing. Together they can spread good seed of improved varieties and contribute to the longer term objective of a more systematic seed production and marketing program. As programs develop and some of the "farmer–seed multipliers" become "salesmen," the more conventional methods of seed processing, packaging, and distribution can be used.

Michael Harrison, who participated in the introduction of new varieties of maize in Kenya, has commented that "the main incentive for demonstration plots was that after the harvest demonstration and joint measuring of the yield, the seed ears were up for sale at double normal price, buyers to select their own ears." Control of the demonstration plots permitted some control of seed production; program officials did not endorse any demonstration that was not up to standard. There was great competition among the farmers to have demonstration plots and thus become seed producers and sellers. As a consequence, demonstration plots were well managed.

In India, acceptance of new wheat varieties was accelerated by the National Demonstration Program of the Indian Council of Agricultural Research. Recognizing the difficulties of convincing farmers with results obtained on experiment stations, the scientists set up demonstrations in farmers' fields. Usually they selected the poorer farmers in the village to participate in the demonstrations so that the success of the new technology could not be dismissed as the effects of affluence.

In the Philippines, to increase rice and maize production, local meetings, short courses, and workshops have been used not only to teach farmers about the new varieties, but also to teach landlords, input suppliers, bankers and credit agencies, civic leaders, and even farm radio broadcasters about them.

Early in its experience with IR8, IRRI gave two kilograms of seed of the new variety to any farmer who visited the institute, but made no public announcement. Within a few months, people from every province of the Philippines had obtained packets of the seed. This is remarkable given the difficulties of transportation and communication in a country made up of thousands of islands. In an area where a single crop is highly important in the lives of the people, informal communication channels work rapidly even though the message sometimes is distorted.

A production rice-kit program launched in the Philippines in 1967 contributed significantly to the rapid adoption and spread of IR8 and its associated cultural practices. The rice kit, first introduced by the government in cooperation with the U.S. Agency for International Development, sold for approximately twenty U.S. dollars. Each kit contained sufficient inputs to plant 2,000 square meters: six kilograms of seed, a few kilograms of insecticide and rat poison, forty-four kilograms of nitrogen fertilizer, and an instruction booklet written in the local language. Farmers bought 22,000 kits the first year, and the success led two commercial fertilizer companies to package and market their own kits. Other kits were assembled and sold to farm youth groups. A rice kit enabled the small farmer to minimize possible financial and yield losses while testing the performance of the new technology, and to evaluate the risk associated with adopting it on a larger scale. With the yield from his kit-planted plot, he was able to plant several hectares with the new variety the following season.

Creating Public Interest

Even though agricultural programs are intended to serve the public, they must compete for such scarce public resources as money, physical facilities (warehouses, ports, heavy equipment), supplies (seed, fertilizer, fuel, spare parts), and professional manpower, including experienced managers. Consequently, it is important for campaign leaders and those responsible for public relations and publicity to plan how to gain the understanding and support of various segments of the public. Through newspapers, magazines, radio, television, posters, and leaflets, people can be informed about seed of a new variety, its importance, and how it can be obtained. It will be advantageous to hold field days, demonstrations, and exhibits for clubs and other groups and to make speakers available to organizations serving farmers and consumers such as schools and professional associations. In countries where the need to increase production is great, a nationwide campaign involving public and private sectors may be warranted. In many countries, companies and civic groups sponsor yield contests to stimulate interest among the farmers and the general public. The range of techniques is limited only by the imagination and resources of those interested. Some governments hire public relations firms or advertising companies for assistance.

Staff and Resources

Differences in scale, organizational arrangements, and other circumstances make it impossible to discuss the staff and resources required for a crop production campaign other than in general terms.

Kinds of Staff Workers and Types of Competence

Seed programs that have reached stages 3 or 4 can call on seed technologists to assist in the educational process, in addition to the national coordinator, the national crop production board, and the staffs of the extension service and crop research and development program. For example, seed certification technologists should not only do quality control work but should join crop production specialists to provide training programs for seed growers, persons with seed enterprises, and farmers. More advanced programs could include seed technologists as subject-matter specialists whose primary responsibility would be to work with seed growers, seed enterprises, marketing groups, seed sellers, and extension personnel.

Crop production specialists, field agronomists, and seed technologists who work with farmers on seed and new varieties should possess technical competence (knowledge of the subject matter), scientific competence (how to obtain or validate knowledge or carry out field trials), economic competence (how to make decisions and recommendations in the light of economic and management relationships such as cost/benefit ratios), farming competence (the skills required to grow the particular crop), and communication competence (how to teach others). Few individuals have all these types of competence, however, so it is important to take them into account when selecting persons to form a team to work in an area. In addition, if administrative and supervisory personnel themselves possess these types of competence, they are more likely to appreciate their value when establishing the teams.

Facilities, Equipment, and Other Resources

Certain activities require special facilities and equipment in addition to what is needed to produce and process seed for trials. For example, if an organization trains agronomists, the special needs include classroom and meeting space, storage for seed and supplies, work space for assembling kits and processing harvested crops, fields for practical training, production implements and tools, instructional aids such as blackboards and projectors, and access to duplication or reproduction equipment.

The field agronomist's work with farmers calls for access to laboratories for basic analyses of seed, soil, and water, for diagnosis of diseases, and for identification of insects. The work also requires transportation facilities to move supplies and personnel.

Overall, the success of the crop improvement and production program will depend upon an adequate budget for salaries, operational and maintenance costs, and travel expenses. Also important are personnel policies and incentives that encourage the staff to strive for greater achievements.

SEED MARKETING

To influence national agricultural productivity, good seed of superior varieties must be purchased and used by thousands of farmers, so seed marketing focuses upon the user (farmer) rather than the product (seed). Seed marketing involves continuous and systematic determination of consumer needs, accumulation of the seed and services to satisfy these needs, communication of information to and from potential consumers about the seed and services, and distribution of the seed to the consumers.

The inability of leaders to appreciate the differences between the technical requirements for varietal development, production, and regulation and those for marketing is a major cause of failure in seed marketing. A marketing program requires personnel with specialized training and expertise not normally found among agricultural technicians responsible for other aspects of the seed program. Marketing personnel require knowledge of human relations, communication, marketing techniques, logistics, and business management. The organizational

---

Essentials of Selling Seed

Attract potential buyers
With seed available, appeal to buyer's need
Establish direct contact with the buyer
Use knowledge of the improved variety to make the sale
Help buyer recognize the value of good seed of improved
   varieties
Give the buyer as much service as possible

---

structure and method of operation are different from those of programs that inform and teach farmers. Thus, seed enterprises need to develop their own marketing programs. Marketing groups are also needed, such as farmers' cooperatives; fertilizer, pesticide, herbicide, and farm equipment suppliers; banks and credit suppliers; and buyers of farm products. Seed enterprises and marketing groups each need to identify and train personnel in seed marketing.

Determination of Consumer Needs—Marketing Research

The perceived demand for seed is not the same as the actual demand. Staff members of a seed program often perceive a demand for seed without having sufficient information. In the absence of a realistic assessment of demand, far too much or far too little seed may be produced.

Market research is the systematic gathering of information concerning the consumers' needs, desires, and buying habits; the number of potential consumers with the needed buying power; and the alternatives available to potential consumers. The information gathered can be analyzed to predict current and future needs of the consumers and the organizational activities necessary to meet their needs.

The purpose of market research is to establish realistic goals or sales estimates. Planning for accumulation of stocks, communication and promotion, means of distribution, and the budget for each activity must be related to the goals. The primary goal of each organization marketing seed might be expressed, "To sell X tons of seed of varieties A, B, and C."

Market Demand

Assessing market demand is not easy. Market demand is the total volume of a product that will be bought by consumers using specific technology in a defined location, within a specified period, and with a certain marketing effort. In this definition, *product* refers to the specific item for which the demand is determined—crop, variety, and quality level of the seed. For example, the demand for all seed is different from the demand for Certified Seed. *Bought* refers to the desires of the consumer who will pay for the seed. In marketing, nothing happens until the consumer buys the seed. *Technology* dictates that the concept of demand relates to the situation in which the seed will be used. For example, a change in the demand for seed of a variety may occur because fertilizer prices change. *Location* imparts a geographic dimension to the concept of demand and implies that demand for a product will vary geographically. *Period* relates to the length of time that an effective demand will exist. For instance, rainfall patterns often limit the planting period. The dates for planting some improved varieties or new crops may differ from those of traditional varieties or crops. *Marketing effort* recognizes that demand can be influenced by promotional campaigns, distribution efforts, and price. Practically stated, people only buy a variety of seed when they know the variety exists, when seed is available, when they have the resources to pay for the seed, and when they believe use of the variety will benefit their operations.

The Demand Forecast

Forecasting market demand for individual seed enterprises is based on what people say, what people do, and what people have done. One method of forecasting demand from what people say is to ask the buyer. This method is practical for a farmer–seed grower who sells to his neighbors, other retail seed sellers, and seed wholesalers.

A second method of assessing demand from what people say is to ask each seller what he believes can be sold in his locality. Seed sellers know more about real market conditions than any other group. The estimates of seed sellers can be separated readily into kinds and varieties of seed wanted, the time they are needed, the area of use, customer characteristics, and the level of marketing effort needed—the components of market demand. Because the marketing personnel of a seed enterprise participate in the forecasting process, they will have greater confidence in the estimates, and thus their incentive to make the estimates realities will be increased. During the early stages of a seed program's development, this technique will be successful to the extent that marketing personnel are willing to cooperate, that estimates are unbiased or can be corrected, and that the personnel involved recognize the need for forecasts.

The components of seed marketing

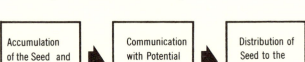

The "expert opinion" method, a third approach to demand forecasting from what people say, works only when the marketing activities and organizations specializing in public opinion surveys are well developed. It is essentially a poll of informed people. It uses few facts, places little responsibility on the estimator, and is most reliable when used for forecasting total needs rather than needs of individual operations. It is suitable for determining trends in crop area and yield levels, but unfortunately it is the most common method used for establishing specific goals in new seed programs.

Determining what people do is the most useful technique for estimating a demand for services. This method measures buyer reaction under actual marketing conditions. For example, market tests may be conducted to determine consumer preference for container size, the effects of size grading, or the acceptance of seed treatment materials enclosed in bags of seed.

Determining what people have done is based primarily upon historical data. It is the way established seed enterprises usually forecast demand, but it is not possible for a new seed program to use this method until background data are accumulated. One technique for making forecasts from data of past performance is a time-series analysis. In a time-series analysis, it is assumed that the actions of people in the past indicate what they will do in the future. Two weaknesses of this technique are the scarcity of historical data and the high probability of change when new varieties are being introduced.

Another technique based on historical data is a statistical demand analysis, which attempts to determine the direct relationship between use and the components of market demand. This technique can identify the relationships among the various demand factors—for example, the effect that introducing a new variety at a higher price will have on the use of a current variety. This method is centered around the statistical technique called regression analysis. Advantages of this method are that it will predict the probability of occurrence—for example, the probability of sales of a variety being five thousand kilograms—and it will provide the probable error from the estimate—for example, total sales of the variety may vary by one thousand kilograms. One limitation of this forecasting tool is that the information concerning each demand factor must be reliable and quantifiable. Also, an electronic computer is nearly always needed to process the large amounts of data involved. Usually this method is better suited to national program projections than to the needs of an individual seed enterprise or marketing group.

Use of the Demand Forecast

Everyone responsible for producing or marketing seeds needs reasonably accurate forecasts of the seed demand, variety by variety. The demand forecast should permit the following questions to be answered specifically: What varieties to produce? Where to produce? How much to produce? When to produce? Where to store? How much to store? What to promote? How to promote? When to distribute? How to distribute? At what price to sell?

During the first five to ten years of a seed operation, demand forecasts are most accurately determined by what people say. People selling seed informally and continually survey the consumers' buying intentions and accumulate stocks to supply their needs. Such independent seed sellers who have several sources of supply can be effective by forecasting their needs a few days or weeks ahead of the actual sale.

On the other hand, seed marketing groups and individuals that are closely linked to seed production must decide how much to grow one to two years before sales will be made. If such long-range forecasts are based on what people say, they are generally less accurate and have a high risk; thus many organizations that sell in several locations use more than one method of forecasting and systematically make the demand forecast in order to minimize risks.

Accurate demand estimates are important to retailers competing for an increasing share of the seed market. A gross misestimate of demand may drive a retailer out of business. When a public or private monopoly exists, demand estimates get less emphasis and this often results in less efficient operations.

But problems do arise. For example, Mexico has supplied wheat seed to a large number of countries. Frequently, the demand was not known until a crop failed somewhere and large supplies of seed were needed. It is difficult to plan, produce, and store seed to meet all emergencies.

In the early years of a seed program's development, seed production should receive priority since seed must be available before it can be marketed. But a major responsibility of seed enterprise leaders is to determine how rapidly to shift the emphasis toward marketing. Market research is an effective tool when making this decision.

Accumulation of Seed Supplies

The second major function of seed marketing is to determine how to get enough Certified Seed or commercial seed to meet the estimated demand. Seed may be produced by or for the seed enterprise to which a marketing group belongs, or seed may be produced by suppliers, inside or outside the country, not associated with the marketing group. Sometimes all sources of supply may be called on.

Enterprises That Produce and Market Seed

Seed enterprises with any of the organizational patterns discussed in Chapter 4 could operate marketing programs of their own. Each of them would have seed production and marketing units. The accumulation of stocks would rest primarily with the production section. However, the marketing section would communicate anticipated requirements (kind, variety, quantity, and quality of seed needed) to the production section far enough in advance of the growing season to permit the seed to be produced and prepared for marketing. As the seed is made ready for marketing, the production section would inform the marketing section of the exact quantity and quality of each variety available.

To make this system function, the managers must make sure there is effective internal communication between the production and marketing sections. This is a particularly difficult task when seed is being grown and stored at several locations.

Separate Production and Marketing Organizations

Some sellers of seed may be independent of production organizations. Examples are farmer–seed sellers, farm supply organizations, government institutions or agencies primarily concerned with activities other than seed supply, and wholesale seed distributors. These sellers may obtain seed from private seed growers, government farms, seed enterprises that sell seed wholesale, or suppliers outside the country.

In a free marketing situation various formal and informal links often develop between production and marketing groups. Generally, this stimulates suppliers to produce seed efficiently and marketing groups to purchase and sell at prices attractive to a buyer. Sometimes, government farms or public enterprises control all production, but seed is marketed through all available channels. The reverse is also possible, with all or most seed growing being done by private farms but with marketing under government control.

The situations in which seed imports may be a factor in the seed supply chain are discussed in Chapter 4. Importation is a normal function of marketing, and a seed enterprise or any other marketing group might sell imported seed. When a government undertakes massive imports, it is prudent to use established marketing channels for distribution and sale.

Administrators and managers must foster communication between producing and marketing groups. One valuable function of wholesale seed distributors and seed brokers is that they provide this service to both groups.

Market Communications

Seed marketing groups need to be in touch with the seed consumer. High quality seed of a superior variety may be priced right and distributed properly and nevertheless may fail to sell well because communication with potential buyers is ineffective. Seed enterprises and seed marketing groups must establish their names or trademarks with buyers. Even when a company trademark or a variety is well known, persuasive communication is essential to stimulate sales, to build the enthusiasm of salesmen and dealers, and to encourage the farmers to buy. Market communications can be divided into four activities: promotion, public relations, selling, and dealer development.

Promotion

Promotion is the stimulation of demand. Specialized personnel are needed who can translate technical information into practical terms for potential customers to encourage them to buy seed. The creative use of promotional materials and publicity is vital to successful marketing.

Publicizing a brand name or trademark, the superior qualities of the seed, and the locations at which seed can be obtained are common types of promotion that stimulate seed sales. An organization large enough to have dealers usually shares in the dealers' publicity efforts by supplying materials they can adapt or by sharing the costs of promotions initiated by the dealers. Techniques such as mini-kits and production kits, radio use, and publications, discussed earlier in this chapter, are also applicable in seed marketing.

The promotional work of seed enterprises and marketing groups that sell varieties developed by public research groups should complement the latter's research and educational activities. Informational material prepared by the public agencies can be used by all.

Administrators and marketing managers concerned with promotion and publicity approve promotional plans, provide the budget for implementation, and promote the organization and its products.

Public Relations

Public relations are activities designed to create a favorable impression of the organization and the individuals who work for it. Some studies show that more than 80 percent of an organization's total sales are attributable to the organization's reputation, employee morale, and public confidence; technical attributes account for the remaining 20 percent.

Administrative personnel, especially, play a critical role in building and maintaining good public relations. In successful organizations, administrative policies concerning public relations are established and rigidly enforced. Employees should be honest, courteous, and friendly and use efficient procedures. Support of service and community improvement projects, informal meetings with public officials and business leaders, and a cooperative attitude in meeting various requests build good public relations. Cleanliness and maintenance of facilities and surrounding grounds help to establish a favorable image.

Selling the Seed

The ultimate objective of all activities in a seed program is to get the seed used. There is a significant difference between selling seed and having seed for sale. Selling is actively searching for a buyer, convincing him of the value of the seed, and exchanging seed for money or other goods. Successful selling can be observed in any village market. The six essential characteristics of selling are used by each vendor who (1) attracts potential buyers, (2) appeals to the buyer's need for the product he has available, (3) establishes direct contact (in this example, verbal), (4) makes the sale through knowledge of the product, (5) helps the buyer recognize the product's value, and (6) gives the buyer as much service as possible, perhaps including credit. Marketing managers should stress these characteristics of selling with sales personnel and dealers.

Efforts to attract buyers should include information about what seed is available, where it can be purchased, and, usually, why it is best purchased from a particular seed seller. The presence of the seed within the transport range of the buyer ensures availability. Distribution is discussed in detail later in this chapter because a lack of availability is a major impediment to increased seed use in most new seed programs. As the proverb says, "You cannot sell from an empty basket." In Kenya, getting improved maize seed within walking distance of tens of thousands of small farmers posed a problem for the Kenya Seed Company. But the

The growth of education, information, and marketing components

GETTING SEED USED	Stage 1	Stage 2	Stage 3	Stage 4
Education and information	*	**	****	********
Seed spread from farmer to farmer	*	**	**	**
Seed marketing through seed enterprises and marketing groups		**	****	********

company found some answers by observing the distribution techniques of the beer and cigarette companies, which had their products in small shops in every village. The breweries and tobacco companies attributed their success to no credit, a big markup (30 percent), and rapid turnover. Consequently the seed company selected two or three shopkeepers in each village to encourage competition, but not more or there would have been insufficient incentive. When possible, new dealers were selected by driving along a road, finding a good field of maize, and tracing it back to the shopkeeper who sold the seed. Forty wholesale distributorships were established to make it easy for each shopkeeper to replenish his seed stock quickly. To obtain a commission, shopkeepers had to buy seed in quantities of at least ten 10-kilogram sacks. In the weeks before planting, many shopkeepers turned their stock over two or three times a week. Seed company representatives visited the wholesale and retail distributors at planting time and kept supplies moving.

Direct contact permits a farmer to establish what degree of confidence he will place in the seller. Consumers rarely purchase seed without talking to the seller. Farmers often prefer to buy seed from a relative or neighbor rather than from someone they do not know, even if it means a sacrifice in value. For this reason, many seed enterprises use local people, often leading farmers, as sales representatives. A buyer's need to know the salesman often works to the disadvantage of city-reared government employees involved in the sale of seed. The technician's economic and social interests are frequently not associated with those of the rural people they are trying to serve.

A seed salesman who does not know how each variety performs under local conditions will make few sales. Since consumers cannot readily examine the two most important characteristics of seed, viability and genetic potential, they will not know the full value and quality of the seed until it is planted and harvested. If the seed lacked the qualities hoped for, it is too late. Thus, the seed seller and the marketing manager should attend crop demonstration field days, consult research and extension personnel, and ask users about the advantages and disadvantages of each variety.

The price and value of a bag of seed are in reality the contrasting opinions of the seller and buyer. Price is the seller's concept of the value of the seed and services offered. Value is the buyer's perception of the benefits to be realized in exchange for money or goods. When a farmer perceives that the value exceeds the price, a sale is made.

Values that a farmer can derive from using good seed of superior varieties include increased knowledge about the purity and germination of the seed, better stands, less contaminating weed seed, reduced susceptibility to insects and diseases, and increased yield potential. These values can be "sold." Many planners believe that the "poor, traditional farmer" cannot pay for "high-priced" seed—and that seed must be given away or sold at a subsidized price. While farmers often save seed or buy from their neighbors, many farmers in numerous countries have bought seed of high-yielding cereal varieties costing two to ten times as much as seed of the traditional varieties, demonstrating that farmers understand price-value or cost-benefit ratios regardless of their level of income or education. When competition for sales exists, value is also enhanced by the seed seller's availability if the buyer is dissatisfied, by services before and after the sale, by association with persons the farmer trusts, by the technical competence of the seed enterprise, and by the willingness of the enterprise to replace seed without charge when the enterprise is at fault.

Dealer Development

Seed enterprises and wholesalers generally market seed through dealers. Prospective dealers should be selected for their effectiveness in the local community and their fiscal responsibility. Identifying dealers who are handling other inputs is a common approach.

The responsibilities a marketing manager assigns to seed dealers are to sell seed actively; to write orders for seed in advance of the planting season; to receive seed from the supplier and keep it in good condition; to arrange for the customer to collect the seed at a designated location (unless the dealer is selling from a vehicle); to complete the sale and collect the money when a buyer receives the seed; to provide service to buyers throughout the growing season; to inform the supplier year-round concerning local conditions, customs, and complaints and compliments about the seed and services; and to pay for the seed according to the policy agreed upon.

To help dealers, a seed supplier provides advertising and promotional materials; prepares technical information about the varieties for sale and their seed quality; conducts training sessions to provide sales personnel with technical information about each variety and guides for effective selling; sends specialists, on request, to investigate major complaints; informs dealers of government programs that influence sales such as credit for farmers and farm suppliers, educational and promotional activities, and research and development; provides order books, price lists, and related operational materials and forms; develops and maintains sensible policies concerning credit, seed-return privileges, pricing, and related matters; refers inquiries to dealers for follow-up in their sales areas; and avoids underpricing the dealers.

Distribution

Distribution completes the process that converts the physical and biological properties of seed produced to economic value for the seller. The distribution system is so important that it can directly affect the goals of a crop breeding program. For example, distribution difficulties have caused some maize breeding programs to change from developing hybrids to concentrating on open-pollinated varieties. Hybrid maize seed may be distributed to as many farmers in the first year as seed of a nonhybrid variety (see Figure 17). In the second year, however, seed of varieties, that, unlike hybrid seed, does not have to be produced annually, can spread from farmer to farmer, whereas new hybrid seed must be distributed to each farmer each year. Several nations that have developed strong marketing and distribution systems, however, use hybrids extensively. A good distribution system must also be operating for effective initial and continued sales of varieties that are not hybrids. Distribution needs to be considered in terms of marketing channels and logistic functions.

Marketing Channels

Seed passes from the producer to the user through a marketing channel. Producers can reach prospective customers directly, through retail dealers, or indirectly, through accumulator wholesalers who in turn distribute to one or more intermediate wholesalers or retailers (see Figure 18). Especially when many outlets are necessary, producers often find it advantageous to market seed through a chain of intermediaries rather than directly because intermediaries carry some of the financial load of distribution and they expand the skills, experience, efficiency, and consumer contacts needed in marketing. Intermediaries may market seed more economically than the producer because of their scale of opera-

Figure 17. Seed distribution patterns for hybrids and other varieties

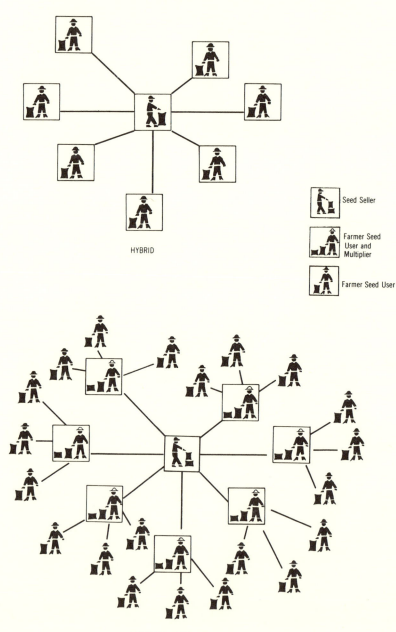

HYBRID

Seed Seller

Farmer Seed
User and
Multiplier

Farmer Seed User

VARIETIES — NOT HYBRIDS

Figure 18. Marketing channels for seed

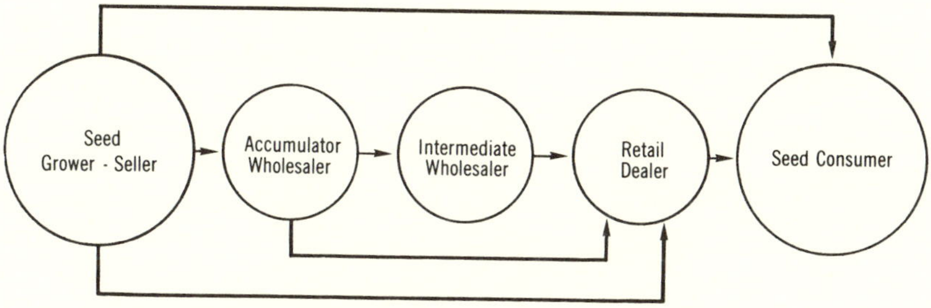

tions, knowledge of local needs, and customer contacts, so both producer and consumer benefit.

A widespread use of seed is most often achieved by using decentralized or multitiered distribution channels. Because of their proximity to the consumer, wholesalers and retailers are in the best position to determine the consumers' reactions to the seed and services offered. They can render services to the customer before and after the seed is sold and pass on valuable information to the seed producer.

Seed characteristics such as total volume, perishability, bulk, unit value, and customer-service requirements have an important bearing on the marketing channel used. For example, a farmer could produce one ton of rice seed and sell it all to his neighbors, while another farmer who produced a ton of onion seed would have far more than his neighbors needed. He would have to depend upon many others to distribute the seed. Compared with rice, most buyers require much smaller quantities of onion seed as it has a higher unit value and much less is needed to plant a hectare. Thus, it takes a more extensive marketing channel to reach enough buyers to absorb a ton of onion seed.

The channel or channels chosen should complement the promotional and communication efforts of a seed enterprise, act as a feedback mechanism for establishing market demand, and serve the seller and buyer at the least cost to both.

Logistics

Such activities as packaging, storage, inventory, transport, and handling are called logistics. They ensure that enough seed of the desired varieties is moved from the place of production to the place of sale so it is available when required. Other examples of logistics are customer services, financing during storage and movement, information systems, insurance against losses, invoicing, and collection.

Logistical expenses add little value to seed, but through them a seed's economic value is realized by the seed enterprise. The cost of the logistics of seed marketing is affected by (1) seed production and consumption features such as climate—which may hasten the deterioriation of both seed and packaging materials—seasonality of production and use, and spread of the production and marketing areas; (2) infrastructure including transportation, accumulation points, and storage facilities for locally produced and imported seed, especially in rural areas; and (3) socioeconomic conditions as manifested in handling skills and the performance of equipment—which influence delays, damage, and losses during transport—in the availability of credit in rural areas, and in the vigor of markets for crops.

One-third to one-half of the price a consumer pays for most agricultural seed involves costs incurred after the seed is produced and processed. Many of these costs result from logistical activities. In developing countries the share of expenditures for logistical activities in seed programs tends to be higher than in developed countries. One obvious way administrators and marketing managers can reduce logistical costs is to avoid production of more seed than can be marketed. Maintaining inventories is the most costly single logistical expense because of the perishability of seed. Moreover, marketing expenses are directly related to size of inventory. To maximize efficiency and thus reduce costs, inventory levels must be lowered or marketing areas must be selected more carefully. However, decreasing inventory levels to reduce costs may also reduce the number of farmers that can be served. Data from market research can help the administrator and marketing manager decide how to strike a balance between the number of people served and inventory costs.

Another way to reduce logistical costs is by customer coordination. Farmers and retailers can be encouraged to place more advance orders. The handling system can be improved by shipping seed at the same time as other inputs. Stock can be shifted more carefully to avoid over- or under-stocking. When one organization controls all levels of the distribution system, it is tempting to order that such changes take place. But ordering customers—independent retailers and farmers—to take a certain action is rarely effective and often damaging. Before changing an established pattern of distribution, many customers' views should be considered.

An efficient internal information system that allows customers' orders to be anticipated and filled promptly can lower logistical costs. Where the order originated, what it contains, and when it should be filled all set this part of the logistics system in motion. An internal information system should be designed to rapidly obtain market information from customer orders, to start the flow of information when the order is taken, and to give priority to actions required to fill the order.

Transportation is another major expense of marketing seed. Transport costs are directly affected by the distance, quantity, means of transport, and kind of seed involved. Shipping distances can be shortened by increasing the number of production areas, processing centers, or storage warehouses, but the costs of facilities and labor will rise. Farmers and dealers differ greatly in accessibility: seed can reach some by truck or animal power, but for many farmers seed will have to be transported on people's heads or backs. Transportation costs can cause the price to be prohibitive for some farmers. In these situations, varieties should be used that do not have to be replaced often, and farmers should be taught how to multiply and save their own good quality seed.

Large seed operations often can reduce costs by increased mechanization in warehouses. This may be socially unpopular because jobs are needed, but it is readily justified economically. Other factors that may decrease marketing costs are improved operations, competition, and increased off-season use of labor to perform certain tasks.

Pricing

The price that a farmer will pay for seed is determined by his perception of its benefit to him and his ability to pay. The price established by a seller includes all direct costs of production and marketing, profit (if any), and an estimate of what the buyer will pay. Pricing is the way in which an administrator can most delicately influence a selling organization's effectiveness in attaining its goals.

Farmers are suspicious of something provided free or at low cost. Free or heavily subsidized seed will not lead, in the long run, to a good seed program because incentives are lacking and no opportunities occur to accumulate capital to invest in seed enterprises. Some of the most successful seed programs have developed where seed was sold at a price high enough to cover all the costs of production plus a profit for the seed enterprise and seed seller. In Kenya, hybrid seed was four times the grain price for many years without meeting the demand. Since the early 1960s both public and private seed enterprises in India have priced hybrid and nonhybrid seed at a rate high enough to ensure a profit.

The price of each bag of seed is based primarily upon four components: direct costs, indirect costs, returns above total cost, and marketing factors such as the total supply of seed available, the price farmers receive for their produce, the availability of production credit, the actual market demand for seed of the varieties to be sold, and the degree of competition among sellers.

Direct Costs

The direct costs of producing and marketing Basic, Certified, or commercial seed are essentially the same, however the percentage of the total cost associated with the direct costs of each category differs, sometimes significantly. The principal direct costs and their relative importance in pricing nonhybrid cereal seeds are (1) production—seed stock, normal crop production, roguing, quality control, and the grower's premium (in contract seed production); (2) processing—drying, cleaning and grading, bags, tags, treatment, and storage; and (3) marketing—market analysis, accumulation of stock, services and promotion, distribution, and risk of loss.

Indirect Costs

Indirect costs are the expenses of operation that are incurred without direct regard to the class of seed or volume of sales—such as taxes, insurance, building maintenance and repair, and depreciation. Also included are regulated expenses that arise from the operation of an enterprise and are controllable within limits by the management—salaries of permanent employees, advertising, heating and cooling, lighting, and standby power. All of these expenses occur regardless of whether the enterprise sells a hundred tons of seed or a thousand. In general, private enterprises and farmers operate with lower indirect costs than do comparable public operations. Many government enterprises do not include indirect costs when pricing seed because these expenses are paid for from government revenues or, as in the case of taxes, the government enterprises are exempt.

Returns Above Total Costs

Individuals and organizations that have sufficient funds to invest in a seed enterprise normally have other investment opportunities as well. To attract enough money to cover the direct and indirect costs, as well as the costs of land, buildings, and facilities needed before operations begin, a seed enterprise must offer investors a return on their prospective investment that is at least equal to the return on alternative uses for their money. For example, the minimum return above total costs, or profit, to investors should be equal to the interest paid by bank savings accounts. Generally investors expect an even greater return than that because there is a greater risk of losing money.

When a return on investment capital is anticipated, the price level of seed must reflect the relationships between all costs and the user benefits regardless of whether a seed enterprise is operating in the private or public sector. Thus, seed of a new or distinctly superior variety may be priced significantly higher than seed of an older variety, because of the increase in value as perceived by the consumer, even though the direct and indirect costs per unit of seed for both varieties may be nearly equal. This method of establishing the price level permits seed enterprises to offer a wider range of seed and services than, for example, when price ceilings are based upon the price of grain.

Government seed enterprises may operate on a profit-paying, break-even, or subsidy basis depending upon the policies of the government. The policy selected directly affects the price of seed. For example, government enterprises that do not have to pay interest on invested funds or on short-term loans can sell seed at a lower price than private businesses can. However, more and more government seed enterprises are being operated on a profit-paying basis—the profits being paid to the national treasury or used to expand the enterprise.

Managers of seed enterprises, both public and private, need to be able to determine the level of sales or production that will provide sufficient income to cover all costs. The amount of profit or loss depends upon how far sales are above or below the break-even point (Appendix F discusses the use of the break-even chart as an overall management control device).

Marketing Factors

Marketing factors may be defined as the interaction between the seller (supply) and the consumer (demand). The major components a selling seed enterprise contributes to this interaction are the quantity and quality of the kind and variety of seed marketed in relation to the demand and the per unit marketing costs (direct and indirect costs plus profit). The major components a farmer-consumer contributes to this interaction are the perceived average price of the seed plus or minus a differential influenced by the kind of seed, the money available to the farmer, his location in relation to the market, his evaluation of the quality of the seed, the value of the supplementary services, and the alternatives available for obtaining seed.

The major alternative available to a farmer who grows "basic" food crops is the opportunity to save the regenerative part of the plant, which may be considered either as grain or seed. Thus, to this farmer, the seed price is equated with the grain market price, which does not take into account direct and indirect costs or the return on investment. Thus, price competition for seed of the "basic" food crops is much keener in most countries than that for seed of hybrid varieties, forage species, vegetables, or flowers because sellers compete not only with one another, but also with the option a farmer has to save his own seed.

Stable seed prices attract private participation in a comprehensive seed program. Seed that cannot be readily saved by a farmer—because of environmental conditions, because the crop is consumed before seed maturity (many forages, fruits, and vegetables), or because of a genetic loss of productivity (hybrids)—is less subject to user competition and the wide price fluctuations common to seasonal markets for farm produce.

GOVERNMENT POLICY FOR GETTING SEED USED

The role of government in influencing acceptance and use of good seed of improved varieties, in educating farmers and others about seed, and in seed marketing can range from being a promoter to being the primary agent. Administrators in government can stimulate the use of good seed of im-

---

Getting Seed of Improved Varieties Used: Major Policy Points

1. Kind and intensity of educational and informational programs and campaigns to get more seed of improved varieties used
2. Measures required to connect research, extension, and seed supply groups for harmonious and efficient work
3. Role to be played by seed enterprises and other groups in the marketing of seeds
4. Kind of assistance the government will give to stimulate the formation and growth of marketing groups
5. Role of government in projecting seed needs and determining demand for seed
6. Desirability of linking the marketing of seed with marketing of other agricultural supplies, and the availability of credit for the entire system
7. Whether seed prices should reflect all costs of production plus a profit to the seed enterprises and marketing groups, or be controlled, or be subsidized
8. Kind of plans to be developed and procedures to be followed to meet emergency seed supply needs

proved varieties by establishing goals for the adoption of new high-yielding varieties and for the use of seed of these varieties. This step can be especially useful when it is linked to helping a commercial seed industry grow through close cooperation with the government in reaching these goals. To ensure that national targets are meaningful, not merely theoretical, administrators at the national level should facilitate market research beforehand.

Informing, advising, and teaching farmers and others about seed may involve many groups. However, the primary responsibility normally rests with the government administrator and specialists responsible for agricultural education and information.

Seed marketing, on the other hand, is a decentralized activity, potentially involving many people outside the government. The government should establish a clear policy on the role it will assume, and it should serve as a catalyst for more effective marketing by groups outside the government. Governments can help marketing groups become established or more effective. They can establish special programs to train and motivate marketing managers and seed sellers to improve performance. Government policies can make it easy to call on foreign experts and seed enterprises to train local personnel and strengthen marketing activities. Even when a government totally controls the marketing channel, marketing principles should be applied.

Government administrators can also help the seed program by making sure that adequate credit is available at all levels. In government farm credit programs, linking credit to the use of inputs can help expand the use of good seed. In addition, however, credit is needed for the organizations involved in marketing (see Chapter 4).

Emergency seed supplies usually are a government concern. In countries where emergencies occur frequently, administrators need to prepare, for example, by producing seed between normal seasons, by making arrangements to obtain supplies of seed from neighboring countries, or by maintaining an emergency seed reserve.

REFERENCES

Andersen, Per-Pinstrup, and Diaz, Rafael O. 1975. A Suggested Method of Improving the Information Base for Establishing Priorities in Cassava Research. In *The International Exchange and Testing of Cassava Germ Plasm*, ed. Barry Nestel and Reginald MacIntyre, pp. 51–60. Ottawa: International Development Research Centre.

Barker, Randolph, and Anden, Teresa. 1975. Factors Influencing the Use of

Modern Rice Technology in the Study Areas. In *Changes in Rice Farming in Selected Areas of Asia*, pp. 17–40. Los Banos, Philippines: International Rice Research Institute.

Byrnes, Francis C. 1975. The Role of Communication in Agricultural Development. In *Readings in Development Communication*, ed. Juan F. Jamias, pp. 69–80. College, Laguna, Philippines: University of the Philippines at Los Banos.

Byrnes, Francis C., and Byrnes, Kerry J. 1971. Agricultural Extension and Education in Developing Countries. In *Rural Development in a Changing World*, ed. Raanan Weitz, pp. 326–351. Cambridge, Mass.: MIT Press.

Foster, T. H. 1972. *Market Measurement and Forecasting—With Special Reference to the Indian Seed Industry*. Mississippi State, Miss.: Seed Technology Laboratory, Mississippi State University.

Gerhart, John A. 1975. *The Diffusion of Hybrid Maize in Western Kenya, Abridged by CIMMYT*. Mexico City: Centro Internacional de Mejoramiento de Maíz y Trigo.

Golden, William G., Jr. 1974. Sri Lanka's Agricultural Extension Service. In *Strategies for Agricultural Education in Developing Countries*. New York: Rockefeller Foundation.

Law, A. G.; Gregg, B. R.; Young, P. B.; and Chetty P. R. 1971. *Seed Marketing*. New Delhi: National Seeds Corporation and USAID.

Mager, Robert F. 1975. *Preparing Instructional Objectives*. 2d ed. Belmont, Calif.: Fearon-Pitman.

Myren, Delbert T. 1970. The Rockefeller Foundation Program in Corn and Wheat in Mexico. In *Subsistence Agriculture and Economic Development*, ed. Clifton R. Wharton, Jr., pp. 438–452. Chicago: Aldine Publishing Co.

Rogers, Everett M., and Shoemaker, Floyd F. 1971. *Communication of Innovations*. 2d ed. New York: Free Press.

U.S. Department of Agriculture. 1954. *Yearbook of Agriculture—Marketing*. Washington, D.C.: U.S. Government Printing Office.

Warriar, R. N. 1972. *Logistics of Fertilizer Marketing*. New Delhi: Fertilizer Association of India.

Wortman, Sterling, and Cummings, Ralph W., Jr. 1978. *To Feed This World: The Challenge and the Strategy*. Baltimore: Johns Hopkins University Press.

7

Personnel Development and Staffing

Planning and implementing a successful seed program is impossible without skilled and motivated people. A lack of qualified staff members is the major barrier to establishing sound seed programs in developing countries. The seed program administrator can use this chapter as a guide to personnel policies, staff requirements, leadership development, personnel management, and training.

COMMITMENT AND POLICIES

A seed program can be properly staffed only if the government considers the seed program integral to the development of the agricultural sector and the national economy and is committed to keeping qualified personnel in the program. Frequent changes in leaders and other personnel can cripple a seed program. A government should help institutions involved in the seed program coordinate their personnel needs to avoid counterproductive competition for scarce qualified personnel.

High-level government administrators can assist the seed program by properly implementing government policy in establishing operational guidelines and making personnel assignments. Potentially successful programs may die despite sound policies if the administrators fail to carry out the policies conscientiously.

Department and division heads should be vitally concerned with properly training and motivating their staffs. Good managers, planners, and specialized technicians are hard to obtain and keep. If well-qualified persons with a potential for leadership or technical capabilities are hired, the program will be on a solid footing.

The managerial, technical, and nontechnical staff members of a seed program must feel responsible to the total program in order for operations to be efficient and effective. The commitment of all these staff members to the seed program is shown by their efforts to improve their capabilities, their diligence in fulfilling their duties, and their loyalty.

Once commitment to the seed program is firmly established, a policy for personnel development and staffing must be set. Policy is the expression of a government's position on a specific subject. A clear and consistent personnel policy provides a strong basis for establishing priorities and planning operations. But policies should be dynamic and have some degree of flexibility. Changes are often necessary.

Although policies for personnel development and staffing must fit the social and economic patterns of a country, some general principles can be stated.

1. All levels of personnel should be covered—from high-level managerial, technical, and administrative personnel to nontechnical workers.

People are at the center but needed are . . .

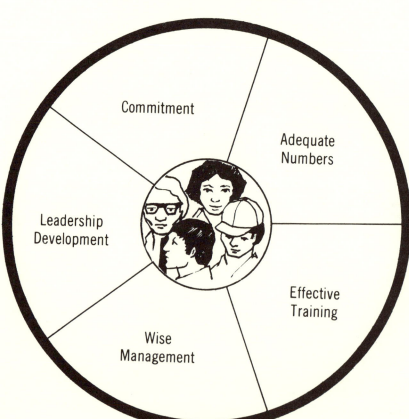

2. Persons trained in seed technology and seed enterprise management should be hired and properly utilized, especially at the decision-making levels.

3. Guidelines for promotion and salary increase should establish one scale of classification based upon years of experience and job performance and a second scale based upon level of training. (Salary increases should reflect the cost-of-living and inflation indexes.)

4. Provisions should be made that would permit technical and nontechnical personnel to shift among government seed activities and between government activities and private seed organizations.

5. Staff members should be selected and promoted on merit rather than on political or other criteria.

STAFF REQUIREMENTS

Staffing should be considered in relation to the program's stage of development, the responsibility (decision level) of the position, and the components of the program. In advanced seed programs the need for program planners imposes special personnel requirements.

Staffing in Relation to Program Growth

The timing and order of a seed program's development must influence administrators when recruiting and training personnel. Even countries that have no organized seed program probably have some of the components of one. Most countries have some type of agronomic research under way. Varieties are being improved by plant breeding or by introduction from research programs abroad. Experience in these activities is appropriate for seed workers, but administrators will have to evaluate whether available personnel are qualified or need additional training.

Research organizations engaged in plant breeding and testing often start seed programs. In this situation, called stage 1 (see Table 4; also Chapter 1), the plant breeder multiplies a small quantity of seed of a variety superior to those commonly in use and distributes the increase to interested farmers. The original seed may have been developed in the plant breeding nursery, imported, or collected within the country and distributed through extension agents, schools, or individuals. In this stage the "program" requires no extra personnel; extra effort on the part of interested people is sufficient.

In stage 2, the research department assigns a staff member to the seed project. This person is responsible for multiplying the seed from the breeding program and organizing distribution. Collaborating with the plant breeder, the staff member employs proper production and handling

Table 4.

Personnel development and staffing in relation to program growth

Stages in Seed Program Development		Types of Personnel Required
Stage 1	Seed multiplication within the plant breeding department; distribution by whatever means available	In a small program, possibly no personnel needed in addition to the breeding staff
Stage 2	Increased volume multiplied as in the first stage, but distribution to selected farmers and seed multipliers handled more systematically	A technical staff member concentrating on seed multiplication and distribution
Stage 3	Develop national policy under which a seed program will be planned; implement programs in seed production, marketing, quality control, seed certification, and training	Policy makers, planners, administrators, managers, and, when the program gets under way, qualified personnel will be needed as seed producers, processors, marketers, designers of seed facilities, seed certification and testing technologists, equipment operators, and trainers
Stage 4	Special attention given to development and strengthening of commercial segment of seed production and marketing; improve cooperation among extension agencies, education and research institutions, farmer and seed groups, equipment manufacturers, credit agencies, and others; reexamine national policy, establish and implement seed law; training continued	Research workers, extension personnel, teachers, farmers, seedsmen, bankers, engineers, equipment operators, seed technologists, newsmen, planners, and managers; number of personnel may be small or very large depending upon size and scope of the program; regardless of number of people involved, the talent and skills required are many

techniques to maintain varietal purity and seed quality. Farmers are then selected to develop a second generation of multiplication; these farmers later may become good seed producers. The number of personnel required during stage 2 will depend upon the number of crops handled, the quantities of seed produced, facilities and equipment, and distribution methods. Such a program probably should never require more than one technical person assisted by nontechnical workers.

When a multiplication and distribution program reaches the point where the research department does not have the space, land, or facilities to accommodate further growth, serious thought must be given to expanding outside the research area. It is time to plan a complete seed program containing all of the components necessary for a seed industry to develop and grow. This is stage 3.

Early in this stage, decision makers and planners must define national policy, examine or develop organizational structures, and identify physical and human resources. These persons must understand the basics of agricultural development and some should have had experience with the seed program and the commercial seed industry. If none have had such experience, a consultant should be called in. After policies have been defined and general planning has been completed, personnel to implement the program must be hired. It takes time to find or train personnel for managerial and technical responsibilities in seed production, processing, marketing, and quality control. Depending upon the availability of personnel, this stage of seed program development may be short or long.

In stage 4, the program develops into a full-fledged seed industry or remains a small ineffectual project. The commercial aspects of production, marketing, financing, and incentives must be given more attention. A seed law to regulate marketing will be needed.

Also in this stage, more emphasis is placed on ties with organizations and people who can influence the progress of the seed program—extension service personnel, news media, farmers' organizations, equipment manufacturers, educational institutions, research programs, and others. The national policy should be reviewed and plans reformulated if necessary. If the program is developing as it should, there will be a ready market for people whose skills have been developed on the job and through training offered by the program.

Staffing in Relation to Decision Levels

The seed program is a complex of interrelated, highly specialized activities that require both technical and administrative leadership at various levels. Table 5 relates the types of decisions—policy formulation, planning/interpretation, and implementation—to a hierarchy of decision makers. Normally, policies are formulated by political leaders while interpretation and implementation are the responsibilities, in varying degrees, of high-level technical officials, managers, administrators, and technical and nontechnical workers. People at decision levels 2 and 3 are responsible for the flow of ideas and instructions throughout the various components of the seed program. Since a seed program has many interrelated activities, the leaders of the components of the program need to work together. The people at levels 2 and 3 are the keys to the success of this working relationship, and the adequacy of their leadership will directly influence the functioning of the entire seed program. The staff members at levels 4, 5, and 6 are primarily responsible for implementing policies and plans. They too must work together cooperatively.

A seed program requires many staff members who have special qualifications and training. Table 6 classifies decision levels in relation to the nature of activities, kinds of decisions and actions, level of supervision or coordination, qualifications, and training. This table can be used as a guide for establishing staff positions and the qualifications needed to fill them.

Because a total seed program includes various organizational units, Table 7 has been developed to relate decision-making levels to the kinds of positions normally found within these organizations. Not all programs include all organizations or all positions; however, the activities connected with these positions are found in most mature programs. Some individuals may have responsibility for more than one position. Thus, Table 7 can serve as a guide to the kinds of positions needed at different decision levels.

Staffing for Program Planning

When the seed activities have reached stage 3 (see Table 4), the seed program should be planned more systematically. The planning of a seed program must reflect the basic policy guidelines. Planning is the interpretation of policy, and failure at this stage will undermine the whole program. Implementation, on the other hand, is the scheduling of inputs to guarantee that outputs are produced according to plan.

The assignment of people to lay plans, to make decisions, and to lead others is never an easy task. Requirements for the size and position of a planning team will vary with a country's strategy for development, its political and geographic characteristics, the size and technological level of its seed program, the resources and people available, and the need for external assistance.

A planning team needs decision makers who have the administrative and technical backgrounds to ensure thorough planning of each component to provide adequately for implementation of the plan, to identify managers and technical people to carry out the plan, and to ensure a balanced development of all components of the program. The planning coordinator or team leader should know agricultural planning and finance. Some experience abroad will be helpful in establishing contacts with international development banks and technical assistance agencies.

Table 5.
Decision level in relation to position

Decision Level	Position	Kind of Decision
	Policy Formulation	
1	Political leaders: president or head of state; minister or secretary of state	Normally government officials who are the leaders of the agricultural sector of the country have the power to make policy decisions, to analyze and evaluate them, and to reformulate, if necessary; to have the desired impact on national development, projected policies must receive sustained support
	Planning/Interpretation	
2	High-level officials: general manager, general coordinator and chairman of the board; general secretary and executive secretary	Planning and interpretation decisions should always be made within the limits established by the formulated policy; all action at lower levels takes direction from clear planning and interpretation of policies
3	Director of department or department head; division head	
	Implementation	
4	Branch manager; head of Basic Seed unit; experiment station head; seed testing laboratory head	With good planning and interpretation, implementation is easier, but no amount of planning can substitute for good implementation; people responsible for these stages are primarily concerned with *how*
5	Other technical people such as field agronomist; seed certification, testing, and law enforcement technologists; processing plant operator; seed producer; seed grower; seed seller	
6	Nontechnical people such as laborers	

Table 6.
Personnel development and staffing: Characteristics and
requirements in relation to decision level

| Characteristic and Requirement | Decision Level | |
	1. Political Leaders	2. High-Level Officials
Nature of activities	Extremely complex and difficult; frequent unexpected situations require masterful strategy and tactics; time consuming	Complex, needing continuous evaluation in order not to be affected by unpredictable events; time consuming
Kinds of decisions and actions	Make most important decisions; responsible for first and basic actions; formulate policies	High-level decisions related to policy planning and interpretation
Level of supervision or coordination	Direct and coordinate policies	No coordination required; provide coordination and supervision
Qualifications	Ability to think strategically; administrative talent; broad political, economic, and social knowledge	Good knowledge of country and commerce; willingness to delegate authority; high leadership capacity
Training	High-level training; international experience helpful	Graduate training; international experience desirable

Decision Level			
3. Directors of Divisions; Department Heads	4. Technical Managers	5. Other Technical Workers	6. Nontechnical Workers
Certain levels of complexity and catalytic effects; key point of interpretation of policies	Complex to noncomplex	Noncomplex but having characteristics of constant innovations	Simple and routine but often requiring careful work
High-level decision making and actions directly related to implementation of the program	Related to procedures for implementation; actions directly affect activities of the program	No policy decisions; work follows set procedures; can use own criteria to interpret procedures	Decisions only related to specific skills
Need some supervision; provide coordination and supervision	Need some supervision and coordination; provide supervision	Need frequent supervision; provide some supervision	Under continuous supervision; do not provide supervision
Creativeness; ability to establish continuous innovation for policy interpretation	Ability to allow formal and informal distribution of authority; professional and technical abilities; administrative talent	Specific technical qualifications for the field of work	For certain activities specific skills needed
College degree, preferably graduate training; special leadership and seed training	College degree, graduate training useful; special leadership and seed training	College degree for highly technical staff, medium-level technical training for others; special seed training	Secondary level for specialized laborers preferable; short courses for certain jobs

Table 7.
Personnel development and staffing: Organizational structure in relation to the decision level and position (not all positions are necessary in all programs)

Component	Decision Level					
	1. Political Leaders	2. High-Level Officials	3. Directors of Divisions; Department Heads	4. Technical Managers	5. Other Technical Workers	6. Nontechnical Workers
Crop Research and Development Program*	...	General research coordinator	Department head; division head; crop research coordinator	Research station head	Plant breeder; pathologist; entomologist; seed technologist; research assistant	Specialized laborer
Basic Seed Program	...	General manager	Department head; division head	Basic Seed unit head	Field agronomist; seed technologist; processing plant operator; seed grower	Specialized laborer; nonspecialized laborer
Seed Production and Marketing Organization	...	Chairman of the board; general manager	Department head; division head	Branch manager	Field agronomist; processing plant operator; seed technologist; seed grower; seed seller	Specialized laborer; nonspecialized laborer

Seed Certification Program	...	Seed certification board; executive secretary; deputy secretary	Department head; division head	Certifying agency manager	Seed certification technologist; seed testing technologist	Specialized laborer
Seed Law Enforcement Program	...	Deputy secretary; deputy director general	Department head; division head	Seed law enforcement head; seed testing laboratory head	Seed law enforcement; technologist; seed testing technologist	Specialized laborer
Extension Program*	...	General extension coordinator	Department head; division head; communication specialist	Crop production specialist (regional); information specialist	Crop production specialist; field agronomist	Extension assistant

*Crop Research and Development Program and Extension Program have been included to show the close relationships with the major components of the seed program. (If testing is to be the primary focus, the staff numbers and qualifications can be reduced.) Training is not included because it may involve all components and be organized under any of them.

The leader must be able to guide and encourage the team members.

Skilled seed specialists, agronomists, and economists will be needed to evaluate a country's technical, social, political, and economic situation as it relates to the seed program. In addition, they can prepare guidelines for seed production and seed processing units, cost-of-production studies, and seed testing laboratory development as well as analyze schemes, identify special seed production areas, and make strategy recommendations. As the planning progresses, personnel such as extension specialists, crop research specialists, and representatives of the commercial seed industry should participate.

Besides professional qualifications, a team member should have tact, a pleasing and dynamic personality, the ability to establish friendly relations with professionals in international and domestic organizations, the temperament for working on a team, the ability to delegate responsibility, and the willingness to operate within established policies.

LEADERSHIP DEVELOPMENT

Since seed technology is new in many countries, leaders often need to be developed or recruited from other programs. The term "leaders" includes all persons in a decision-making or management role. Leaders at the various levels make different kinds of decisions. Depending upon the decision-making level, the leader's role may encompass program development, general or specific internal management, or political contacts. All organizations need leaders who are technically and administratively competent, who can handle responsibility, and who can make changes effectively. There are other characteristics to consider when identifying and recruiting leaders for seed activities:

Ability and personality traits. The leader of a seed program needs organizational ability and must be able to work smoothly with subordinates and with leaders of other components both inside and outside of government. Tactful leadership should induce key staff members and colleagues to identify their interests with the welfare of the whole seed program. Without this quality a leader is ineffective and may be a liability.

Potential leaders probably will lack some technical skills, but these can be added by training, and, as A. T. Mosher observes, "It is therefore better to select those characteristics that are more difficult to change by training."

Interest and motivation. The interest and motivation to work effectively both in and outside the office are vital for maintaining the enthusiasm of a staff. Although leaders of seed activities must do some paperwork, much time should be spent outside the office. Persons without these qualifications should not be considered for leadership positions in seed certification, seed law enforcement, Basic Seed multiplication, seed enterprises, or seed supply organizations.

Creativity and innovativeness. A leader frequently must be creative enough to find ingenious solutions to knotty problems. He also must be sufficiently innovative to reach beyond the routine, especially in a rapidly changing situation. Thus, creativity and innovativeness are important characteristics to be considered when selecting a leader.

Willingness to delegate responsibility. A leader must recognize the need and have the willingness to delegate responsibilities to others on the staff.

PERSONNEL MANAGEMENT

Clear personnel policies and creative personnel management can make a staff more effective. When new components of a seed program are being started, opportunities often exist to apply innovative personnel management techniques. Worthwhile objectives are improved morale, higher motivation, longer tenure of service, increased professionalism, and better performance. Several measures can promote these objectives:

Organizational structure. Many seed activities must be systematically performed. The organizational structure can contribute significantly to the way work is done and to maintaining clear communications among personnel. Keeping the lines of communication and channels for action open creates the atmosphere of stability required for high morale and good staff performance.

Job classification for seed technologists. Since seed technology is new in many countries, it may not be properly classified. In order to gain a promotion, seed technologists may have to compete for positions in plant breeding, agronomy, or other fields. A proper job classification for seed technologists will help keep trained and experienced technologists in seed activities, contribute to staff morale, and build professionalism.

Job description. A good job description specifies the responsibilities and level of authority associated with a position. It helps prevent misunderstandings between the administration and the staff member. (Appendix A includes a job description for a seed maintenance and multiplication technologist.)

Delegating responsibility. In seed activities, so many details must be handled correctly and on schedule that all decisions cannot be made by one person. A leader should judiciously delegate some responsibilities but retain top priority responsibilities. Excessive delegation of responsibilities can insulate top leaders from personal contact with progress and major problems. As responsibility is delegated, a system of communication between the leader and those to whom responsibility is given must be created through frequent personal contact, regular staff meetings, and visits to field activities.

Effective use of staff. People should be placed in positions suited to their abilities, skills, and training. For example, individuals with good dexterity and patience perform well as seed testing technologists. A person who likes to meet people, understands farmers, and enjoys the out-of-doors will be a better seed certification technologist.

Incentives. People like to feel appreciated and useful. Promotions and proper compensation for tasks well done fortify that feeling. Private and semiautonomous seed enterprises usually have greater freedom to build staff morale through monetary compensation than do government organizations. Some organizations provide credit to staff members for purchasing a vehicle, if one is needed on the job. This step, when combined with a travel allowance, can be an incentive for better performance.

Recognition and praise. Commendation can be a powerful morale builder and an effective force to achieve more output per person. Yet how frequently this is ignored or forgotten. As A. T. Mosher puts it, "A good administrator seldom suffers as a result of building up the reputation of his subordinates. On the contrary, he builds staff morale *and* his own reputation as an administrator when he gives his subordinates full credit for what they do." Employees should be inspired and motivated rather than driven. It is often effective to ask for advice from a subordinate, thereby allowing him to develop his own "order." Although this ability must be developed through experience, the morale of most staff members can be kept higher when they play a part in making the decisions and "orders" are held to a minimum.

Even when directives must be issued, the way they are given can affect morale. A very special part of each person, including the one who sweeps the chaff off the floor, is self-respect. Leaders in programs need to care about their staff members as individuals and to show them that this concern is felt. Causing work to move forward while maintaining the self-respect of employees is a skill leaders must develop to keep morale high.

Staff motivation. A common adage is, ability plus motivation equals performance. An effective leader ensures that ability and motivation are combined to ensure performance. In addition to the points already stressed, a leader should appeal to what people want. Too often leaders are so engrossed in what they themselves want that they forget to consider what others want.

Staff morale will be high when employees feel that their jobs are important and worthwhile. Since seed is so basic to life and agriculture, jobs in seed activities do have a significant meaning. Employees need to recognize the importance of their organization to the total seed effort. If their opinion of the organization is good, if they feel it is doing important work, and if they sense that the organization is held in high esteem by others, they will be proud of their work. New employees should be thoroughly oriented when they start work. Staff meetings should be held periodically to review activities, discuss plans, and consider problems faced by the organization. Long-term objectives should be clearly defined so that staff members can see the progress that is being made.

Staff maintenance. The importance of maintaining a trained staff is frequently overlooked. The fact that many skills are needed in a seed program in order to get good seed to farmers does not seem to be fully appreciated. As a result, seed programs rise and then fall as the leadership shifts and the trained staff goes elsewhere.

Although staff members should feel free to seek better jobs, every effort must be made to provide incentives so qualified and experienced personnel will stay. Especially in the public sector, administrators should find ways to remedy weaknesses in promotion systems and salary schedules. Personnel rules *can* be changed if a fundamental problem exists.

Good leaders are constantly looking for leadership potential in their subordinates. Those who are gaining the technical skills should eventually be moved up to leadership positions. Continued training should be offered to staff members so that they have a chance for professional improvement. Developing motivated and well-qualified staff members should have high priority in all seed activities.

TRAINING

Although seed production is similar in many respects to general crop production, the differences require specially trained personnel. In addition to the cultural practices associated with crop production, seed producers or supervisors are concerned with isolating fields, plant spacing and fertilization for seed production, water management, pollination, harvesting time and methods, drying, storage, quality control, distribution of seed, market demands, business management, and finance. Little attention is given to these areas in standard agronomic curricula. The purpose of training is to improve the work performance of personnel. The appropriate type of training depends upon the level of personnel involved; however, broadly speaking, training can be divided into academic and nonacademic types.

Academic Training

Personnel in managerial and technical positions should have college-level training, but many persons who have good leadership qualities hold managerial positions despite the lack of a college degree. The appropriate training for each person will depend upon the requirements of the job to which he will be assigned. Academic training will accomplish little unless the training is relevant to a seed program. Therefore, institutions that have a relevant curriculum should be used whether they are inside the country or outside.

Employment must be coordinated with the training of personnel. Any failure to use new graduates is a disgraceful waste of time, effort, and money.

Graduate Training

While meeting university requirements, the graduate program should fit the needs of the trainee and the position to which he is to return. He should receive special counseling and supervision in his research work, which should be relevant to his country's seed program. Opportunities should be made available outside the classroom to observe advanced seed programs and to study the concepts on which they are built.

Kind and Location of Training

Usually outside the country
  Graduate training leading to a degree
  Undergraduate training leading to a degree
  Short courses for specific advanced study

Inside or outside the country
  Basic short courses
  Conferences and seminars
  In-service training and travel-study programs

Usually inside the country
  Seed appreciation courses

Study and training to the Ph.D. level increases the effectiveness of people to be engaged in plant breeding or research in other areas such as seed physiology, entomology, or seed pathology. Training to the M.S. level is usually sufficient for personnel working in management, seed production, seed processing, quality control, marketing, and other technical areas directly associated with a program. Since many trainees often are assigned additional administrative duties after completing their studies, supplemental administrative courses or practice is desirable.

Undergraduate Training

Only a few educational institutions offer a curriculum emphasizing seed technology that will enable undergraduates to gain sufficient training to be effective in most seed activities. Some institutions have one or two seed courses in the agronomic curriculum or include a special seed emphasis in crop production and plant breeding courses. These courses are valuable for giving a large number of students an appreciation of seed technology in relation to their major interest. Such courses are especially useful for future plant breeders and for people who will work in programs dealing with agricultural education and information.

Nonacademic Training

Not all phases of a seed industry require university training. Short courses, seminars and conferences, in-service training, travel-study opportunities, and appreciation courses can be used to develop the skills of staff members or to upgrade performance.

Short Courses

Well-conducted short courses can be a most effective method of training. When adequate facilities and qualified trainees are available, nearly any type of training can be provided. For maximum benefit to the seed program, training should be available to personnel in every job classification from management to the technical fields of production, processing, and quality control. No person has responsibilities so large or so small that he cannot benefit from a good short course designed to improve performance.

Short courses provide a mechanism for training leaders in basic concepts of the program as well as for developing practical skills at all levels of responsibility. Short courses may be the only training medium used for staff improvement, or they may be supplemental to other forms of training. Some courses may be designed to bring administrators and technical leaders up to date on developments in their fields. Other types of courses may provide basic information to improve subsequent in-service training.

Seminars and Conferences

National seed seminars for policy makers, planners, administrators, and technical personnel focus attention on domestic issues so that recommendations can be formulated for presentation to the proper authorities. Regional or worldwide seminars may be concerned with production, commerce, disease and insect control, quality control, certification, or other problems that may have a bearing on seed programs both nationally and internationally. Such seminars broaden the participants' ideas and often lead to improvements in the operation of a seed program.

In-Service Training

No amount of academic or short-course training will substitute for learning on the job. In-service training can be useful at any level of responsibility. After a short period of work with an experienced manager who has similar responsibilities, new administrative and technical managers may develop an insight into their jobs more quickly than they could without the experience. Where possible, practical training should be provided: persons such as seed certification technologists should be trained in a seed certification program; equipment operators, on a farm or in a seed processing plant; and seed testing technologists, in a seed testing laboratory.

It is not always easy to provide facilities or to locate organizations for in-service training at the time it is required. However, when in-service training is possible, it will help a trainee develop good work habits and will give him confidence in his ability to perform the job.

Travel-Study Programs

Travel-study programs may be of various types. They can be organized for a particular group of trainees to emphasize a specific subject—such as Basic Seed production, seed legislation, seed processing, or seed certification—or to expose a trainee to all aspects of a seed industry. In other instances, a travel-study program may be planned to achieve an objective, and then the trainees who are to participate are selected. Travel-study programs are ordinarily conducted in areas where a full-scale seed industry is established so that organizations involved in different segments of the industry can be visited.

In travel-study programs, participants observe a more advanced seed program in action and learn to use or adapt aspects that may be applicable. Such programs can be strongly influential if the program leaders have clear objectives and provide an effective follow-up. It is possible to have a good program that all participants enjoy but that has little effect upon a seed development program. The success of a program depends principally upon matching the participants in the course and the places visited to the purpose to be achieved.

Appreciation Courses

Conferences or seminars may be useful to acquaint individuals not directly involved in a seed program with the importance of improved seed in agricultural development. Administrators, research workers, extension agents, bankers, agricultural teachers, and leading farmers are examples of people who can be reached through appreciation courses. Their understanding and support can be very helpful in getting a seed program moving.

Training: For Whom and Where?

All personnel need training at some time. The responsibilities of the job and the personal characteristics of each person will determine the most appropriate kind and location of training. Since even people with the same level of responsibility respond differently, matching the needs, responses, and qualifications of the individuals to the training requires skill on the part of the person who selects trainees. The proper selection of trainees to be sent abroad is particularly important. The value of training abroad will depend largely upon a student's proficiency in the language, educational and cultural background, and ability to adjust to unfamiliar conditions. A student's motivation, attitude, health, maturity, study habits, satisfaction (or dissatisfaction) with the course of study, and other cares and worries can also be important. There is no doubt that exposing a student to a different environment broadens perspective. If he has the ability to adapt the knowledge learned to the problems at home, the experience can be quite rewarding, both to himself and to the seed program.

Where the training is done is less important than the quality and relevance of the training. If knowledgeable local people can plan and conduct effective courses, there is little reason to go elsewhere. Some subjects can be learned better within a country than abroad; others can be learned as well in one place as in another. For example, although the general principles of seed production and harvesting can be taught anywhere, specific practices are best learned under the conditions that exist where the crop is grown. On the other hand, if facilities are equal, seed testing and seed processing can be taught and practiced anywhere for use anywhere.

Training at Home

If qualified instruction can be provided, training should be done domestically. Although not all countries can furnish advanced technical training, most can, with some outside help, provide much of the training at the lower levels of responsibility. All nontechnical workers (decision level 6) and most other technical workers (decision level 5) can be successfully trained at home.

Those who plan the training should draw upon whatever resources are available. In some countries these will be meager, in others, numerous. It takes imagination to develop training opportunities. For instance, the ministry of agriculture or education might have departments that can provide facilities and perhaps even staff members for training courses. Sometimes the extension service may be used for training workers. Agricultural universities that are service oriented and have specialists are good choices to provide instruction. Depending upon the kind of seed program and its stage of development, various farm organizations and seed producers themselves offer possibilities for assisting in the training. Other training opportunities are provided by meetings or even short courses, which are sometimes sponsored by farmers' associations, cooperatives, farm service centers, a village agricultural office, or the production or marketing section of a public or private seed enterprise.

If training opportunities are scarce, a seed program can justifiably provide facilities for training or help other institutions provide them. If a university has a good agronomic curriculum at the graduate level, a seed program might encourage the school to develop graduate-level courses in seed technology. The seed program could provide equipment or make grants to employ additional teachers as they become available through advanced training. Or universities that have good undergraduate agronomic programs might be helped to establish seed courses. Undergraduate courses are a good way to introduce students to the profession of seed technology and its importance to agricultural development. These are only a few examples of how seed development activities can ensure that a continuing number of interested and potentially competent people will enter the seed program as it develops into an industry.

Regional Training

Regional programs are useful for types of training not available at home. All technical managers (decision level 4) and some technical workers (decision level 5) will benefit most from regional training. In regional training, compared with other training abroad, transportation costs are lower, language training is less likely to be necessary, and the climate, crops, farming methods, and social customs may be more similar to those in the home country. During such training programs, technicians in neighboring countries become acquainted with one another. Subsequently, friendships may make it easier for countries to exchange technical information and varieties, to set uniform seed quality norms, and to establish interagency certification of seed.

Training programs of several kinds can be provided when the countries that make up a region have moderately well-developed programs. In many instances, regional training will fit the needs of the trainees better than similar studies in countries that have more advanced programs. Some international organizations are willing to help develop regional program plans and assist in training.

Training in Developed Countries

Many students from developing countries receive their academic training in foreign institutions. Some short courses and most travel-study training take place in developed countries. Most international conferences and seminars, as well as short-term courses of various types for administrative and upper-level technical personnel, are held in countries that have advanced seed programs.

Academic training in developed countries is the most expensive and

---

Personnel Development and Staffing: Major Policy Points

1. Whether the level of commitment to the seed program is reflected in the assigning of personnel to the program
2. Whether the role of seed technologist is properly identified in the total scheme for development
3. Whether personnel management procedures improve staff morale, motivation, and effectiveness
4. Whether sufficient funds and personnel are being committed to training
5. Whether trained personnel are being properly utilized or whether changes need to be made to avoid losing valued staff members

most difficult type of training to obtain. It should be restricted to those working at decision level 4 or higher, or to individuals who have the ability to become leaders at those levels. Travel-study programs are normally provided only for leaders at decision level 4 or above.

Training in developed countries provides an opportunity for leaders to become acquainted with the most advanced technology and to observe more mature programs in action. Personnel from countries whose programs are in stages 3 or 4 can gain the most. A major weakness of training in developed countries is the difficulty the trainee has in applying the training and observations to his home situation. Training in a more advanced, neighboring country may make the experience more relevant, especially for people whose seed programs are in stages 1 or 2. (Appendix G provides information on possibilities for special training in seed technology in developed countries.)

Training Materials and Useful References

Any place where training is done should have a library of current books, journals, and magazines that can be used in training and as references. In addition a supply of audiotutorial materials relevant to the training program should be assembled. A large amount of material is available in English, but much can be found in other languages too (see the Bibliography).

Several magazines and other publications containing seed information are published on a regular basis (see Appendix H). Administrators, managers, seed technologists, and personnel in training will find many of these useful.

REFERENCES

Jugenheimer, R. W. 1976. *Corn: Improvement, Seed Production, and Uses.* New York: Wiley.

Mosher, A. T. 1975. *Serving Agriculture as an Administrator.* New York: Agricultural Development Council.

8
Resources

Most seed programs in developing countries face competition for personnel, local funds, and foreign exchange. Seed program administrators must strive to obtain resources when they are needed and to use them efficiently in order to provide good seed in adequate quantities. One of the first questions an administrator must answer is whether all seed is to be produced domestically or whether some seed will be imported. Resources can be conserved by utilizing research from other countries and by drawing on foreign seed production and processing capabilities. For some crops the most efficient use of resources might be to develop a variety testing program, contract for seed production abroad, import the seed in bulk, and develop a means for local packaging and distribution. Under such a system, the best varieties could be supplied for certain crops at attractive prices with a minimum use of local resources.

Public resources can be conserved by encouraging private investment in the seed program. Private investment can accelerate the development of a nation's seed industry. (Ways in which governments can stimulate private investment are described in Chapter 4.)

Whether the public sector carries all the load of seed industry development or only a portion, the administrator must decide what physical resources (such as equipment and facilities), human resources, financial resources, and external resources the program needs and which are available. And he must continually ask, "In what way will this resource provide more good quality seed for our farmers?" A large and costly seed processing plant or an impressive seed testing laboratory does not necessarily guarantee better seed for the farmers. Some seed processing plants process seed but do not improve it; some seed testing laboratories test seed, but no one uses the results. Thus, the seed planted by farmers is unaffected. The administrator's ultimate goal must be to ensure that resources are used to achieve the results intended.

PHYSICAL RESOURCES

Whether to use machinery and equipment in place of manpower is a common and knotty problem for administrators. The level of mechanization may be considered excessive or inadequate. However, the primary issue is, how can seed of improved varieties be multiplied and maintained most satisfactorily? Seed is a valuable product that must have special attention to preserve its quality. This objective should not be overshadowed by a social desire to employ a few more laborers.

In many developing countries manpower is inexpensive however, the possibilities of substituting labor for machines in seed operations are not unlimited. In cleaning large volumes of seed and moving seed from one piece of equipment to another, the speed and accuracy of machines are advantageous. Prompt harvesting and threshing to preserve seed quality are also more easily achieved with machines. Furthermore, training and managing a large labor force is a burden on a management staff, which often is overextended. And as wage levels increase, operating costs for a large labor force may climb at an unpredictable rate.

Much laboratory work, however, requires selective judgment that must be based on human observation. And for tasks such as roguing fields, loading trucks, and some other routine chores, there are no acceptable substitutes for manual laborers.

This section discusses equipment needed for seed growing and harvesting, seed drying, seed processing, seed storage, and quality control. It also deals with obtaining supplies, getting facilities developed, making equipment operative, and maintaining equipment.

Seed Growing and Harvesting

Seed growing and harvesting are specialized operations and require some equipment that may not be common in ordinary farming. For example, uniformity of crop growth is far more important in seed production than in farming because of a need for synchronization of pollination and proper assessment of varietal purity. A precision seed planter that simultaneously applies fertilizer gives far more uniform stands than sowing seed and broadcasting fertilizer by hand. Precision planters place seed and fertilizer properly for quick and uniform germination and optimum use of nutrients. Also, since the two operations are done together, planting takes less time, which is advantageous when weather conditions are unpredictable.

In ordinary farming, hand harvesting and threshing may be quite acceptable, but in seed harvesting great care must be taken to preserve seed quality. For seed of many crops, the proper level of moisture content at harvest is critical and falls within a narrow range, so harvesting must be rapid. Seed quality can also be spoiled by bad weather if the crop is not harvested promptly. Machines reduce these hazards because harvesting can be completed faster. Both plot-sized and commercial-scale field equipment are needed depending upon the stage of seed multiplication.

Plot-Sized Equipment

Equipment for planting small multiplication plots must precisely control the depth, spacing, and rate of seeding. Some equipment is designed for the purpose; other equipment is adapted from standard models. Excess seed should be easy to remove from the planter to avoid carry-over from one plot to another. Planters should be equipped to deposit seed and fertilizers simultaneously.

Depending upon plot size and convenience, either hand-carried or field-sized sprayers and dusters can be used to apply insecticides and fungicides.

Harvesting equipment should be efficient, capacious, timesaving, and easily operated. It must be easy to clean between lots, both to save all harvested seed and to avoid admixture. For harvesting larger plots, commercial units with some minor adaptations are often adequate. Specially designed units may be necessary for harvesting seed of some vegetables and forage crops and when plots are small.

Moisture testers are required to check seed at harvest and during drying. Dryers are often needed to bring seed to the correct moisture levels for storage (see below under "Seed Drying").

Since equipment is frequently moved from one location to another, suitable vehicles or trailers must be available. Transport is also needed to move supplies, tools, and harvested seed.

Addresses of organizations that can provide information about suppliers of small-plot equipment can be found in Appendix H. In addition, the International Association on Mechanization of Field Experiments has a useful handbook that lists many suppliers of plot equipment.

Commercial Field Equipment

What commercial field equipment can be used depends on the size of the farms involved, the concentration of production within an area, and the crops grown. The quantity of Basic Seed, Certified Seed, and commercial seed handled by a seed production unit is also a factor. Equipment used in normal mechanized farming operations is suitable in most seed growing activities. It is beyond the scope of this publication to cover all the different kinds of equipment in detail. Standard seed drills, planters, combines, maize pickers, and similar farm equipment are often satisfactory. Equipment—especially combines—should be chosen for ease of cleaning between lots.

For certain kinds of crop seed, unusual equipment is needed such as detasseling machines for maize and special harvesting machines for some vegetable and forage seed. If the volume of production is not large, manual labor can substitute for this equipment.

Since the use of equipment in seed growing is more important than in ordinary farming, special arrangements may be justified to ensure that the operations of seed growing units, both public and private, are adequately mechanized. For example, seed enterprises may do custom work for seed growers or rent equipment to them for certain tasks. Special loans and import permits could be made available to seed growers to help them obtain equipment. If land leveling is needed to improve seed yields and to facilitate irrigation, specially trained and equipped teams could be formed to provide these services to seed growers for a fee.

Seed Drying

Moisture content influences seed viability enormously. Consequently, artificial drying is almost mandatory for the production of high-quality seed, especially in warm, humid areas. In a warm environment, seed with a high moisture content deteriorates rapidly unless it is dried promptly and properly. Drying must start within a few hours after harvest and continue without interruption until seed moisture has been lowered sufficiently.

Provisions for artificial drying should be made as soon as the production of a substantial quantity of seed is foreseen. The capacity for artificial drying should be matched to other closely related components of the seed system.

There are two compelling reasons for having at least some artificial drying equipment. First, while a few kilograms of seed can be dried in the sun, handling large quantities in this way is difficult if not impossible during rainy periods. Since a concentration of seed production is inevitable and necessary for quality control, artificial drying is required because a small increase in the amount of seed produced easily overwhelms traditional, nonmechanical drying systems.

Second, unless the harvest falls during a period with no rainfall and relative humidities lower than 60 percent, the moisture level of seed in the field will probably dictate a need for drying to preserve viability.

Otherwise, when the seed moisture content is over 14 to 16 percent, biological heating, which is detrimental to seed viability, will occur within the bin. The larger the quantity of seed involved, the more critical are the effectiveness and efficiency of the drying operation to viability.

The drying of seed should be distinguished from the drying of grain. Seed drying is a more specialized, more exacting operation. The drying system must be designed to perform satisfactorily under the climatic conditions that prevail during and after harvest. The most suitable type depends on such factors as the volume of seed to be dried, the number of varieties to be handled, and the size of the seed lots.

The kind of seed should also influence the choice of a dryer. For example, a continuous-flow, tower-type dryer cannot handle ear-maize. Various types of dryers and their advantages and disadvantages in seed drying operations are discussed in *Drying Cereal Grains* (by Brooker, Bakker-Arkema, and Hall).

Although many drying systems are available, devising solutions to drying problems is not simple. Much of the technology has been developed in temperate regions such as Western Europe and the United States. Equipment and techniques designed for drying in these areas cannot be transferred to tropical countries without adaptation to local requirements. But engineering data required for designing seed drying facilities are scarce in tropical environments. Designers are often forced to make projections and guesses and increase safety factors in hopes that a drying facility will operate effectively. It is difficult to dry and store the seed of major crops—maize, sorghum, millet, peanuts, wheat, rice—in the tropics and subtropics, but the problems are even more complex for the seed of oilseed, vegetable, forage, and fiber crops. Although experience in designing seed drying and storage facilities for tropical environments is being accumulated, much remains to be learned.

Seed Processing

The processing plant is one of the largest capital investments in a seed program. Properly equipped and managed, it is a tremendous asset to a seed operation; otherwise, it becomes a major liability. The quality and appearance of the seed marketed are greatly influenced by the different steps in processing: removing contaminants and low quality seed, sizing, treating, and packaging.

All seed used for planting requires some processing. As the production and harvesting systems become more mechanized, seed processing becomes more important. Various contaminants must be removed from "raw seed" to prepare it for marketing and planting. While contaminants such as inert material and off-size seed are not, of themselves, detrimental to production, they do influence ease of planting, insect infestation, and appearance, and they can contribute to storage problems. Contaminants like weed seed, seed of other crops and other varieties, and diseased seed affect production if they are not removed or properly treated. But processing cannot compensate for poor production conditions, so seed growing areas must be carefully selected (see Chapter 4).

Appearance is not often considered an aspect of seed quality, but it affects the promotion and marketing of good seed of improved varieties. A farmer who is planning to buy seed does not see genes for disease resistance, prolific tillering, and other genetic factors; he sees seed that is basically the same shape, size, and color as the seed he has been planting. To convince the cultivator that the seed he is being urged to buy is "improved seed," it must look like improved seed.

Although processing is the principal means of removing contaminants, it can also be a major source of contamination. Processing must be done without contaminating the seed being processed with seed of other varieties and crops. And seed must be protected from mechanical damage during processing. Conveying seed through a series of machines with little attention to the separations effected during each operation is *not* seed processing.

Kind of Equipment

Seed processing requires equipment designed for specific types of separations, other machinery such as conveyors and elevators, and buildings. If large quantities of seed are to be handled, processing should be almost completely mechanized.

Although processing equipment varies in complexity, size, and design, all processing equipment makes separations based on differences in physical properties between desirable material (pure seed) and undesirable material. To satisfactorily remove contaminating material, seed usually must be processed in a specific sequence through several machines, with each machine removing just a portion of the contaminating material.

The choice of a machine or sequence of machines depends on the kind of seed to be processed; the quantity of seed to be handled; the nature, kinds, and quantities of contaminants such as weed seed, other crop seed, and rotten seed to be removed; and the seed quality objectives to be met. Thus, the processor must be as familiar with seed standards and seed characteristics as he is with processing equipment.

During processing, seed must be handled and conveyed mechanically because human labor, regardless of its availability or cost, cannot be an effective substitute. Thus, when determining equipment requirements for processing, such items as conveyors, bins, and elevators must be considered in addition to separation equipment.

When the treatment of seed with chemicals can control serious diseases, it enhances the value of the seed and provides further justification for the farmer to purchase the seed. Some systemic seed treatments are extremely valuable. Consequently, seed treating equipment, to uniformly apply chemicals, is necessary in the seed processing chain.

The use of automatic or semiautomatic bagging equipment at the end of a processing line can ensure that packaging is not a bottleneck. Laborers can do the bagging, but the health hazards of handling chemically treated seed and the problems of keeping several people working at a rate comparable to the rest of the processing line means it is usually preferable to use equipment instead of manual labor. The equipment should be capable of handling small as well as large bags since many kinds of seed should be distributed in small bags.

The Processing Line

The combination and sequence of equipment and operations—the processing line—should be tailored to the essential objectives of processing. Often, a simple processing line consisting of an air-screen cleaner linked with conveyors, surge bins, elevators, a treater, and packaging equipment is entirely satisfactory. Operations that do not contribute to the processing objectives or enhance the real quality of the seed should be avoided. In processing maize seed, for example, it is not necessary to produce several size grades when most of the seed will be planted by hand or with simple planters. Some size grading may be warranted to improve the appearance of the seed, but excessive grading should be avoided.

The movement of seed through a processing facility is frequently illustrated by flow diagrams such as shown in Figure 19. Flow diagrams generally disregard restrictions on resources and usually show all equipment items needed to produce high quality seed—a product that merits a premium price. When resources are limited, only the equipment essential to meet existing seed quality standards should be selected. When a program is just becoming established, few seed facilities require all the equipment shown in most flow diagrams.

Minimizing equipment, however, will increase the importance of good management of operations prior to processing, especially in growing and harvesting the crop. Processing operations can be minimized if the raw seed is of relatively high quality. Thus, emphasis must be placed on producing and harvesting seed under optimum conditions.

Figure 19. Seed processing flow diagrams

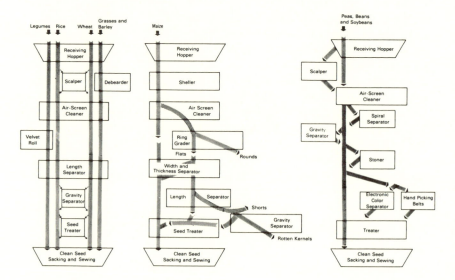

Equipment in Relation to Crops

The basic processing needs for many major crops such as maize, millet, sorghum, rice, wheat, barley, and oats are similar. Generally, the quantity of each of these seed crops is large enough that they all can be processed by the same machines with little difficulty. Even the seed of some forage grasses and forage legumes can be processed by the same equipment used for field crop seed.

Vegetable seed processing, beyond the most basic cleaning operation, is a specialized activity. Combining vegetable seed processing with the processing of field crop seed should be considered with extreme caution even though resources are limited. It is uncommon even in highly developed seed industries.

The seed of certain crops (such as peanuts and cotton) requires such specialized processing equipment that processing is often done in an installation that handles no other crop.

Seed Storage

In most subtropical and tropical regions, seed programs that do not have adequate storage facilities are likely to suffer losses while holding seed stocks from harvest to the next planting. The risk is even greater for stocks held in reserve for longer periods of time. High temperature and high relative humidity lower seed viability rapidly and favor storage insects. Rodents are a constant threat. Scandal may arise if large amounts of seed with low germination must be discarded or, worse, if it is sold to farmers. Thus, administrators must see to it that all components of a seed program have the capacity to store a wide range of seed in varying quantities for long periods of time. Personnel in research and development programs have to store breeding materials and germplasm satisfactorily. Those who multiply and maintain the initial seed increases of new varieties need to store seed for short and long periods (see Chapter 3). Seed enterprises must store large quantities of seed from the harvest season to planting and keep unsold seed or reserve stocks for longer periods. Dealers and distributors are obliged to keep seed viable until it is sold to farmers. Because of these differing requirements, nearly all seed programs need facilities for short-term, intermediate, and long-term storage (see Table 8).

Assessing the Situation

Seed is stored under such differing conditions from country to country that generalizations cannot be made about what is required in one particular location without good information concerning temperature and relative humidity. Information must be gathered on the kinds and amounts of seed to be stored since different crops have different storage requirements. The amount of space for short-term, intermediate, and long-term storage needs to be assessed. The availability of local building materials, insulating materials, electricity, and air-conditioning and dehumidification equipment should be considered. With this information, administrators can plan how to meet seed storage requirements.

Seed Storage Conditions

Lowering the moisture content of seed, and lowering the temperature and relative humidity in which seed is stored, extend the storage life of most seed. The moisture content of seed is directly affected by the relative humidity of the atmosphere around it. Raising the relative humidity increases the seed moisture percentage; lowering the relative humidity reduces the moisture percentage of the seed. Figure 20 illustrates these relationships.

In areas where the temperature and relative humidity are high, seed can lose its viability within weeks. Although a room or building with air-conditioning and dehumidification can be satisfactory for storage, constructing, maintaining, and operating such a facility is costly and

Table 8.
Seed storage needs of various components of the seed industry

	Short Term (6–8 months)	Intermediate (8–20 months)	Long Term (3 years or more)
Crop research and development	Breeding materials selected for current season planting	Breeding materials held in reserve	Germplasm
Initial seed increase	Breeder and Basic Seed for current season	Reserve Breeder and Basic Seed and unused supplies	Breeder Seed of selected varieties, Basic Seed of inbred lines, and special lots of some varieties
Seed enterprises	Seed held from harvest through processing to distribution in current season	Reserve seed and unused supplies	Not normally necessary unless a research program is supported
Marketing agencies	Seed ready for sale	Unused supplies	Not normally needed

sometimes difficult. Therefore, a country should consider other ways to improve storage conditions or reduce storage needs.

Seed might be stored outside a normal seed production area at locations that have lower temperatures (at higher altitudes, for example) or drier climates. Seed might be produced in the off-season in irrigated areas so that the total storage period from harvest to planting is reduced. Growing, harvesting, threshing, and drying conditions could be assessed to see if there are ways to improve the condition of seed as it enters storage. Steps might be taken to improve existing storage facilities with a minimum of investment, for example, by insulating the ceiling and installing an exhaust fan.

Meeting Storage Needs

If an investment must be made in new facilities, administrators should get technical guidance on the design and construction. As illustrated in Table 8, decisions are needed on the amount and kind of storage required for research and development activities, initial seed increases, seed enterprises, and marketing agencies. For seed enterprises, the managers must decide how much space they need for bulk storage and how much for bags; how the amount of storage should relate to the drying, processing, and marketing capacities; how much of the storage capacity, if any, needs a means of temperature or humidity control and the kind of equipment to be used; and what materials are available, locally or abroad, for sealing and insulating walls, ceilings, and floors.

In addition to the storage space, the way seed is packaged must be considered. Indeed, the package is a storage container. Points to consider in packaging are whether the package is porous and will permit air movement or is sealed against air movement; the kinds of materials available for packaging seed; and the ability to dry seed efficiently for totally sealed packaging. Publications such as the FAO's *Cereal Seed Technology* deal with these technical issues in detail.

Quality Control

The sequence of quality control operations in seed activities is discussed in Chapter 5. These measures require four categories of equipment and facilities: equipment for seed certification field inspections, equipment for inspections in processing plants and seed law enforcement, equipment for variety check-plot work, and buildings and equipment for seed testing laboratories.

Figure 20. Seed storage life as affected by the relationships between seed moisture content, relative humidity, and temperature

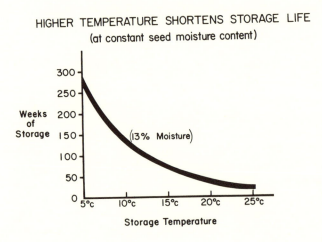

HIGHER TEMPERATURE SHORTENS STORAGE LIFE
(at constant seed moisture content)

(13% Moisture)

Weeks of Storage

Storage Temperature

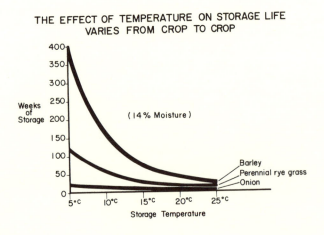

THE EFFECT OF TEMPERATURE ON STORAGE LIFE
VARIES FROM CROP TO CROP

(14% Moisture)

Weeks of Storage

Barley
Perennial rye grass
Onion

Storage Temperature

INCREASING RELATIVE HUMIDITY RAISES SEED MOISTURE CONTENT

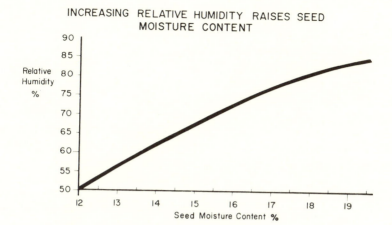

Relative Humidity %

Seed Moisture Content %

HIGHER SEED MOISTURE SHORTENS STORAGE LIFE
(at constant temperature)

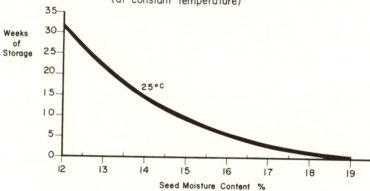

Weeks of Storage

25°C

Seed Moisture Content %

THE HIGHER THE TEMPERATURE AND SEED MOISTURE CONTENT THE SHORTER THE STORAGE LIFE

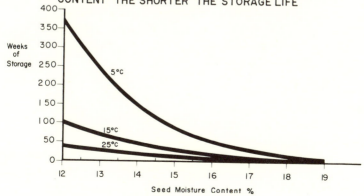

Weeks of Storage

5°C

15°C

25°C

Seed Moisture Content %

Seed Certification Field Inspections

Since field inspections should never be delayed, seed certification technologists must work regardless of the weather, and they may spend long periods of time away from their offices and homes. A means of transportation is essential; it can range from horseback to aircraft. Generally, though, inspectors need transportation for daily travel between farms and fields within a region. Each inspector, or team of inspectors, operating in a region should have a vehicle, preferably one with four-wheel drive. Motorcycles, bicycles, and other means of transportation are sometimes used, but they are less satisfactory.

In addition to a means of transportation, various supplies are needed for field inspections such as stationery, printed forms, tapes, and hand counters. Further details are included in the FAO's *Cereal Seed Technology*.

Inspections in Processing Plants and Seed Law Enforcement

Transportation is the first priority for inspections in processing plants and for enforcing seed laws. Equipment and supplies needed to perform this work properly include sampling triers, portable sample dividers, seals, sealing devices, sample containers or bags, printed forms, labels, and marking pens. Portable moisture testers are also useful, especially in processing plants.

Variety Check-Plot Work

Variety check-plot work deals only with testing for varietal purity and homogeneity. Items such as hand magnifiers and counters that facilitate morphological examination of plant populations, single plants, and single-head progenies are useful. Drills or planters designed to seed single-head rows or plots are indispensable when the variety control work deals with several samples. Single-head threshers and single-ear shellers should also be included on the equipment list. Other useful equipment and supplies include paper-pulp or peat pots for single-plant seedlings and transplantation.

Official Seed Testing Laboratories

Official laboratories are established by law and, as outlined in Chapter 5, are entrusted with testing seed for law enforcement and certification purposes. Such laboratories may also do service testing for seed enterprises, crop research programs, and farmers. To achieve uniformity in test results, they must follow standardized methods and test conditions for which suitable space and certain basic equipment is needed.

At the minimum, a seed testing laboratory must have adequate space for offices and for receiving seed samples, making physical purity analyses, testing germination, and storing seed samples. As a program develops, the laboratory should have, or get access to, productive and uniform land for control plot tests for variety and species, special germination tests, and seed health studies. In some areas and in advanced programs, greenhouse space and growth chambers will also be helpful.

Although many construction details should be the responsibility of an architect, the layout and space requirements should be developed in cooperation with a knowledgeable seed technologist. Details such as insulation and sealing for germination and sample-storage rooms and the location of windows, plumbing, and electrical outlets all need careful attention.

The purity section of a laboratory must have adequate space for microscopes, lenses, working desks, balances, reference seed collections, sample dividers, and temporary storage for working samples. Seed analysts may need special purity-analysis stations to protect them if chemically treated samples are being handled frequently. If possible, some laboratory-size seed cleaning machines should be available to enable the laboratory to advise seed enterprises on processing procedures.

Germination and seed health laboratories need adequate space for working desks to prepare samples and for counting seeds and seedlings; a sink with a water supply; germinators and temperature-regulated germination rooms; storage of dishes, paper, towels, and sand; dish washing; a sterilization oven and an autoclave; incubation cabinets for seed health tests; and equipment for laboratory seed treatment. A place where ovens, mills, desiccators, and balances can be arranged should be provided for moisture determinations.

Space must also be available to unwrap seed packages and register samples as they are received. This should be near the place where results are calculated, the office where reports are typed, and the fee section. The room in which submitted samples are stored should be large enough to hold samples for one year, and in many areas it should be temperature controlled. The head of a laboratory needs a private office. A specialized seed library should be available to the staff.

Some equipment, such as germinators and precision balances, should be ordered from specialized suppliers. More commonplace supplies and general laboratory equipment, however, can often be obtained locally. Simple and timesaving devices—such as seed counting boards, spatulas, sample tins and jars, and plastic trays for germination in sand—can often be made locally, or locally available items can be substituted or adapted. A comprehensive survey on constructing, equipping, and organizing a minimum seed laboratory appears in the *Proceedings of the International Seed Testing Association*, Volume 34, number 1 (Boeke, Oomen, Schoorel, Bekendam, and Koopman). ISTA's *Survey of Equipment and Supplies* contains information contributed by twenty-seven member countries from five continents. The *Seed Testing Manual* by Chalam, Singh, and Douglas, and the FAO's *Cereal Seed Technology* are also useful (see Appendix H for other information about equipment).

Service Seed Laboratories

Service laboratories are usually operated by a seed enterprise and test seed for the enterprise itself as well as for seed growers and farmers. Information about moisture, purity, and germination helps seed growers and seed enterprises as the seed is harvested and during drying, processing, and storage. The services offered are between the laboratory and the user with no official significance. However, the quality of the tests and the results obtained can equal those of official laboratories if the proper equipment is available and the staff is experienced.

A service laboratory uses equipment similar to that of official laboratories, but it normally tests fewer kinds of seed and uses methods that permit faster results. One example of simplification is the equipment used to determine moisture content. Reasonably accurate information obtained from a quick testing device is more valuable during a drying process than a more exact result obtained a day later through a more refined test requiring a grinder, a constant high-temperature oven, a desiccator, a precision balance, and a trained analyst.

Size, Capacity, and Location of Laboratories

The size of a laboratory should be determined by the number of samples to be tested per year or season, the number of seed species to be tested, and, to some extent, the kinds of seed species to be tested. The number of samples that can be tested is affected by the capacity and condition of equipment, the number of seed testing technologists and their skill, the frequency of "difficult-to-test" species, and the time available between harvesting and planting. In addition, the proportion of official samples compared with service samples needs to be considered when determining size and capacity.

Any possible expansion of a seed testing program must be planned too. Laboratories may be established with excess capacity, with provisions for enlarging the building later to accommodate more equipment and staff, or with the intent of starting new laboratories elsewhere as the need arises. Starting new laboratories, however, may make it difficult to keep test methods and interpretation of tests uniform. If mail and other communication services are adequate, enlarging a laboratory is usually better than building a new one.

A certification program also requires facilities for office work, label printing, and storage of equipment and supplies. In many countries, the seed testing laboratory, the seed certifying authority, and the seed law enforcement group are all accommodated in one building. Administrators of these programs should plan their space needs jointly.

Service laboratories are normally small units connected to seed enterprises. Their capacity is mainly determined by the operation of the enterprise and the types of services the laboratory is expected to provide. The best location for a service laboratory is where service is needed—close to a seed processing plant or the office of the enterprise.

Sources of Supplies

Although some of the equipment for the seed industry is unusual, it does not need to be sophisticated. Much equipment can be found locally. The grain industry uses air-screen cleaners, triers, and dividers. Gravity separators are used in the coffee, tea, and mining industries. Dryers, control equipment, conveyors, and elevators are used in many manufacturing and processing operations. Balances, magnifiers, and other items of laboratory equipment are used in biological and pharmaceutical industries. Dehumidifiers are used in the pharmaceutical industry. Air conditioners are commonplace items.

If some equipment cannot be purchased locally, administrators must decide whether to have it made locally or to import it. Local fabrication can be desirable because it saves foreign exchange, uses local labor, makes servicing the equipment easier, and may be inexpensive. On the other hand, considerable effort may be dissipated on numerous items that have a limited potential market locally. Time may be wasted while local fabricators master new skills. And compared with imported items, the quality may be lower and users may have less confidence in the product.

The advantages of importing are that a large selection of equipment is available, much of the equipment has years of evaluation and experience in manufacturing behind it, in the more industrialized countries high standards and competition have a desirable effect on quality and prices, and local resources can be used for the manufacturing of other equipment that may be in greater demand. The disadvantages of importing are that spare parts are hard to find and maintenance may be difficult, especially if several kinds and models of equipment are imported. In addition, imports use foreign exchange and local industry derives no benefit.

In Brazil and India, some seed drying, processing, and laboratory equipment is manufactured, but certain items are still imported. Both countries have large potential markets and a substantial industrial base for manufacturing such equipment. Countries with smaller industrial bases have rarely manufactured any specialized equipment for the seed industry.

Thus, governments and seed program administrators should consider encouraging local manufacturing of seed equipment if the local industry has the ability and if the potential market is large enough. Possible regional opportunities for sales should not be overlooked. If some local fabrication can be undertaken, priorities must be set for items to be made locally and for those to be imported.

To ensure a continuing supply of imported equipment, local dealers might be established and encouraged to carry spare parts and develop service competence. A government can help by providing priority foreign exchange allocations and special loans. When local dealers are not involved and special government imports are made, plans should be laid carefully to avoid frequent major importations. For large and rapidly developing programs, a government might organize a group of individuals who can give the local industries advice about fabricating equipment and who can help seed enterprises meet their equipment and maintenance needs. Governments could also import equipment and lease or rent it to seed enterprises.

One of an administrator's primary responsibilities is to be sure the seed industry has equipment and spare parts. If supplying the equipment necessitates foreign exchange, the necessary allocations should be provided. It is essential that clear decisions be made on how the supply is to be ensured. If a commercial seed industry is to be encouraged, sufficient priority must be placed on the supply of all the physical resources needed to avert shortages that would cause delays.

Developing Facilities and Starting Up

Administrators and leaders in the public and private sectors frequently must erect seed facilities and start them operating. The development stages leading to the operation of a seed drying, processing, and storage unit will be used to illustrate the factors involved and the steps necessary to put such a facility in operating order.

Obtaining Basic Information

When establishing a comprehensive seed unit, drying, processing, and storing must be considered together (although in some situations drying may be unnecessary). After seed has been dried, it may have to be stored for a short while prior to processing. It would be better, however, if the three operations could be carried out simultaneously so the seed can be dried, processed, and moved into storage without delay.

To properly plan capacity and related requirements, the basic information needed about the location includes the

- kinds of crops, the number of varieties, and the month seed of each will be in the facility
- area and yield of each variety and crop
- harvest months and the weather conditions at harvesttime
- harvesting methods and the speed with which harvest is completed
- time needed to move from the field to the processing installation and the kind of transport equipment to be used
- quality of roads in the area
- seed lot sizes to be handled
- expected number of seed growers and seed fields
- days available for drying and processing a particular crop

After this information has been gathered, consideration needs to be given to power supply, rail connections, access by road, and water supply. Administrators can draw heavily upon their technical staffs to gather and evaluate this information. Foreign specialists—from private, governmental, or charitable organizations—may be useful at this stage. Private consultants for seed processing facilities are also available in some countries. When a large facility is planned or several locations are involved, some governments send local administrators and leaders abroad to visit comparable facilities.

Preparing Preliminary Plans

After the basic information has been collected and a site has been selected, preliminary plans or drawings must be made to show the layout of the various components of the facility and the location of equipment. At this point, the size of each item of equipment and, if possible, the specific models must be decided upon.

If a facility is small—a Breeder or Basic Seed unit or a modest private installation, for example—the preliminary drawings may be sufficient for a contractor to proceed with construction plans. For large plants, however, preliminary drawings must be converted, after approval, into final drawings.

Final Drawings

Although the preliminary drawings can be developed by a seed technologist or an agricultural engineer who is thoroughly familiar with drying, processing, and storing seed, the final drawings should be prepared by an architect. Countries with a high degree of industrial development have suitable architects; other countries may have to hire foreign consultants. The architect must fully understand the intent of the preliminary drawings to ensure that the final drawings will meet the requirements. While preparing the drawings, the architect should consult frequently with a knowledgeable seed technologist. Seed flow diagrams will help guarantee that the installation can operate as planned and will be of great help in preparing the architectural drawings.

If the architect's technical background is sufficient, he should simultaneouusly prepare detailed equipment specifications. Otherwise, a seed technologist or an agricultural engineer must work closely with him. Many details regarding space needs, dimensions, and seed flow require careful attention at this point. When preparing equipment specifications a list of spare parts is also needed. The kind and quantity of spare parts can be based upon the recommendations of the probable equipment supplier and others who know the equipment.

Preparation of Documents

The tender documents (documents requesting formal bids to supply goods or services) must list the specifications of equipment and spare parts desired, the construction details for the facilities, and the requirements for installing the equipment. Quality factors should be clearly specified for all facilities and equipment. As tender documents are prepared, consideration must be given to how much of the construction and fabrication can be done by local contractors. Electrical and plumbing installations, the construction of surge bins, and the purchase of ladders may be easily arranged locally. However, major pieces of processing, drying, and seed storage equipment may have to be imported. When equipment is imported, the tender documents should specifically include requests for installation and for operation instructions, service manuals, and complete lists of spare parts. The tender documents should also state to what extent suppliers are expected to train operators and maintenance personnel. Provisions for assistance in operating and maintaining the equipment in the initial month of use should be included. The number of months for training and assistance should be specified, and the amount of follow-up service should be indicated. Documentation should be required for all bids on equipment, installation agreements, and commitments on training and follow-up service, whenever applicable.

When a seed staff is inexperienced, a total tender request that makes one firm responsible for all construction and installation may be most desirable. The cost may seem high, however the responsibility for a complete and properly working installation is more clearly fixed, so the ultimate cost may be less.

Appraising Bids and Awarding Contracts

Because bids are sometimes difficult to appraise, the use of outside assistance may be prudent. Details need to be checked carefully against the original tenders.

Construction and Installation

If an architect is involved, his continued supervision and guidance is essential during construction and installation. Depending upon his background and experience with seed drying, processing, and storage facilities, he may continue to draw heavily upon others who have special seed technology or engineering knowledge.

When the tender documents are prepared, a timetable should be included because plans for seed production depend on the completion date. Construction and installation should be scheduled so that the facility will be ready ahead of a specific harvest season to avoid handling seed in incomplete facilities. From preliminary drawings to a completed installation normally takes twelve to twenty-four months. The many local factors that affect construction time—including the availability of electricity, land, access roads, and transport facilities—need consideration in the early planning stages to prevent unnecessary delays.

Initial Operation

The "start-up" of a seed facility should be a cause for celebration—if care was taken in all of the development stages. On-site training for the staff is extremely valuable though training in a comparable facility may also be helpful. Some equipment dealers are better able to provide on-site training than others. With thorough planning, a facility can be fully operating with a minimum of delay and with a staff competent to continue its operation season after season.

Although emphasis here has been on a seed drying, processing, and storage facility, similar care and precaution is needed in the construction of other facilities such as a seed quality control center or a seed testing laboratory. Additional points to consider when developing such facilities within the framework of a project are given in the FAO's *Improved Seed Production*.

Maintenance of Equipment

The importance of equipment maintenance cannot be overemphasized. Preparations for a maintenance program should start when equipment is being ordered. Manufacturers often can provide additional copies of printed instructions concerning the operation and maintenance of equipment or more detailed information that may not be supplied unless specifically requested. The instructions should be in the language of the requesting country, or they should be translated as soon as they are received.

Equipment maintenance costs money. In preparing budgets, administrators should include funds to cover the importation of spare parts and the cost of local repairs. It should be possible to draw on maintenance funds easily so that prompt action can be taken when it is needed.

Identifying and training people in equipment maintenance is as important as training for any other specialized job. Some suppliers can provide special training programs, at their factory or within the country, on how to operate and maintain equipment. When equipment orders are placed, plans should be made to take advantage of such opportunities. Providing a properly trained maintenance staff with mobility, spare parts, and tools can be one way to help keep equipment operating at peak efficiency in several locations.

HUMAN RESOURCES

A lack of people skilled in seed technology is the major barrier to building sound seed programs. Although special training programs can help (see Chapter 7), there are people in many countries who have been trained in seed technology but who are not being used effectively. Administrators should identify these people and make sure they are in positions appropriate to their talents. In addition, plant breeders, agronomists, and horticulturists can become "seed specialists," with some special training in seed technology. Administrators, however, should avoid taking critical staff members away from other programs.

Physical resources should be in balance with human resources. Investing heavily in physical resources without the proper personnel to operate and maintain the facilities is not productive. Conversely, developing a large staff without having adequate facilities is demoralizing to the staff members.

In this chapter emphasis has been placed on the need for personnel to operate and maintain equipment as a way to stress the importance of human resources to the total program, but many specialists are needed in the seed industry (see Tables 5–7 in Chapter 7) and their qualifications differ.

FINANCIAL RESOURCES

The financial requirements of seed programs differ greatly from country to country, but it is possible to identify some of the major cost items in each segment of a comprehensive seed program (see Table 9).

In addition to using public funds, seed program administrators have two main alternatives for meeting some of these financial needs: the private sector could assume some of the costs, and, as a seed program grows and develops, certain aspects of it can become self-sufficient or nearly so. Consequently it is possible for a program to grow as it moves from stage 1 to stage 4 without a continued expansion of public expenditures (Figure 21). There are several areas in which costs can be shifted to the private sector or to users:

Research and development. Seed enterprises can assume some responsibility for research and development. As seed sales have become more profitable, seedsmen and private enterprises in some countries have provided funds for public crop research and development of special interest to them.

Basic Seed production. When a program is sufficiently advanced to sell Basic Seed to seed enterprises and individuals, most or all expenses of the Basic Seed program can be met through the sale of seed.

Seed enterprises. Whether seed enterprises are private, public, or a combination of the two, their objective should be to become self-sufficient and to provide a profit, even though some public funds may be needed for initial investment and costs.

Seed quality. Seed certification is basically a service provided to seed growing, processing, or marketing groups in the interest of seed consumers. Such programs normally start with full government support, but they can evolve into self-sufficient services. Nevertheless, a government needs to begin such programs with an initial investment and continue to provide some support thereafter.

Seed law enforcement is carried out for the public good. Thus, it is a public responsibility for which a government usually pays the total cost.

Table 9.
Major cost items in a seed program

	Physical Resources	Direct Operating Costs
Research and development	Buildings, land, field, and laboratory equipment	Permanent staff; program expenses such as vehicle operation, equipment maintenance, supplies, and part-time labor
Basic Seed production (as part of research program)	Harvesting, drying, and processing equipment; seed storage facilities	Permanent staff, seed production and processing costs
Basic Seed (as a separate unit)	Harvesting, drying, and processing equipment; seed storage facilities; buildings; land; field and laboratory equipment	Permanent staff, seed production and processing costs, managerial staff, utilities, labor, equipment maintenance
Seed enterprises	Buildings; land; drying, laboratory, processing, and field equipment; seed storage facilities; vehicles	Managerial, technical, and nontechnical staff; working capital; interest on all borrowed capital; seed production costs; transportation; drying; processing; storage; equipment maintenance; seed losses; marketing costs (plus research and development if done by the enterprise)
Seed certification	Building, office furnishings, vehicles, minor equipment such as sampling triers	Staff, field and seed sampling supplies, report forms, vehicle and office operation, utilities, labels, education
Seed law enforcement	Buildings, office furnishings, vehicles, minor equipment such as sampling triers	Staff, seed sampling report forms, supplies, vehicle and office operation, utilities, education
Seed testing	Buildings, laboratory equipment and furnishings, vehicles	Staff, laboratory operating supplies, sample bags, utilities, vehicle operation
Education and promotion	Buildings, office furnishings, vehicles, audiovisual equipment	Staff, vehicle operation, audiovisual supplies, promotional materials, printing and duplicating

Figure 21. Financing a seed program at different stages of development

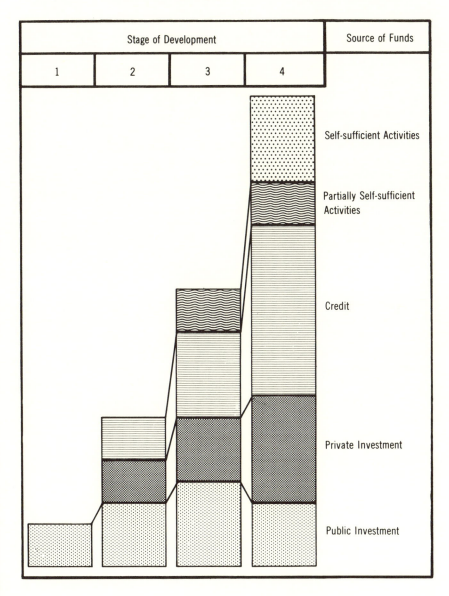

Occasionally, however, revenues are collected from seed sellers to support part of this program.

Seed testing is a service to the seed certification program, to the seed law enforcement activity, and to farmers and other consumers. Eventually some revenue can be expected from the seed testing program, but at first it is better to have free testing to promote the service. Later, some of the costs of seed testing can be covered by charging fees.

Education and promotion. Educational costs are usually borne by a government for the public good. When promotion and sales activities are conducted by seed enterprises and marketing groups, costs are met totally by them.

Government administrators are often interested in the financial benefits to be derived from investment in the seed sector. Since all segments of the seed industry are interrelated, benefits should be considered in terms of the entire seed industry, or perhaps in terms of the total agricultural sector, instead of only one segment of it. Timeliness of operations is vital to the success of every aspect of a seed program. Thus, it is extremely important that funds be available when they are needed.

EXTERNAL ASSISTANCE

A wide variety of external organizations (such as United Nations agencies, bilateral aid agencies, the World Bank and regional development banks, philanthropies, and universities) are able to assist national seed programs through consulting services, grants for equipment purchases, training fellowships, and loans. *Agricultural Assistance Sources,* published by the International Agricultural Development Service, provides information about many of these organizations.

Although the primary responsibility for the development of a seed

---

External Assistance Is Most Effective

When
    high-level commitment exists
    assistance is continued for an extended period
    appropriate assistance is provided at the right time
    a well-conceived project is linked to the total seed program
        and improved varieties

program must rest with each national government, external assistance can be extremely value if

>high government officials are committed to the seed program
>
>an external assistance agency is able to support the activity for an extended time
>
>an external assistance agency can provide well-qualified technical experts to a project at the proper time
>
>a well-conceived project is effectively tied to the total seed program and to meaningful crop research

A publication of the U.S. Agency for International Development *A Guide for Team Leaders in Technical Assistance Projects* points out, "External assistance is itself a scarce resource, whether expressed in funds or in technical assistance personnel. It too is an input for which there may be strong competition." Administrators need to consider not only the possible sources of help but also the most effective ways to use the assistance provided.

External assistance could be used in developing the initial plan for a country's seed program (see Chapter 1). In the process of developing a plan, consideration could be given to the strengths and weaknesses of the local staff and to the adequacy of local funds and foreign exchange. It would then be possible to identify the types of external assistance most

---

Resources: Major Policy Points

1. Whether the allocation of physical, human, financial, and external resources reflects the priority placed on seed
2. Whether investments in equipment are being made systematically with emphasis on achieving good seed quality or whether merely the least expensive or most sophisticated machines are being chosen
3. Whether the importance of equipment maintenance is being emphasized and reflected in the long life of machines
4. Whether credit policies should favor seed enterprises and marketing groups
5. Whether funds are available when needed in all seed operations
6. Position regarding external assistance to support development of a seed program

complementary to the resources available within the country. External assistance should not be accepted just because it is offered. It should be fitted to needs expressed in an overall plan for development of a seed program. It might support an entire program or only a part of it. It might be short-term or long-term assistance. Some programs need only financial assistance, but many require technical assistance to complement financial assistance. Assistance from private businesses might be valuable too (see Chapter 4).

External assistance often could be used more effectively than it is. Administrators should

> clearly state the responsibilities of, and the role to be played by, the foreign expert
>
> assign one or more staff members to work with the foreign expert and benefit from his presence
>
> develop clear plans or project outlines jointly with the personnel of the external assistance agency and the local staff
>
> provide sufficient local support such as office space, transportation, housing, clerical support, and interpretation and translation services to free the foreign expert from excessive concern about details of this kind
>
> periodically review progress to ensure that all local and external groups are working toward the same goals

As stressed repeatedly throughout this section, the resources available are important, but a wise use of what is available is even more important.

REFERENCES

Boeke, J. E.; Oomen, W. W.; Schoorel, A. F.; Bekendam, J.; and Koopman, M.J.F. 1969. Project Seed Laboratory 5000. *Proceedings of the International Seed Testing Association, Equipment Number* 34:115–168.

Brooker, D. B.; Bakker-Arkema, F. W.; and Hall, C. W. 1974. *Drying Cereal Grains.* Westport, Conn.: Avi Publishing Co.

Chalam, G. V.; Singh, Amir; and Douglas, J. E. 1967. *Seed Testing Manual.* Delhi: Indian Council of Agricultural Research and USAID.

Feistritzer, W. P., ed. 1975. *Cereal Seed Technology.* Rome: FAO.

Feistritzer, W. P., and Kelly, A. F., eds. 1978. *Improved Seed Production.* Rome: FAO.

Gregg, B. R.; Law, A. G.; Virdi, S. S.; and Balis, J. S. 1970. *Seed Processing.* New Delhi: U.S. Agency for International Development.

Harrington, J. F., and Douglas, J. E. 1970. *Seed Storage and Packaging—Appli-*

cations for India, New Delhi: National Seeds Corporation and The Rockefeller Foundation.

International Agricultural Development Service. 1979. *Agricultural Assistance Sources*. 2d ed. New York.

International Association on Mechanization of Field Experiments. 1972. *The International Handbook on Mechanization of Field Experiments*. Aas, Norway.

International Seed Testing Association. 1973. *Survey of Equipment and Supplies*. Zurich.

Kreyger, J. 1972. *Drying and Storing Grains, Seeds, and Pulses in Temperate Climates*. Publikatie 205. Wageningen, Netherlands: Institute for Storage and Processing of Agricultural Produce.

Mehta, Y. R.; Gregg, B. R.; Douglas, J. E.; Balis, J. S.; Joshi, M. S.; Rekhi, S. S.; and Young, P. B. 1972. *Field Inspection Manual*. New Delhi: National Seeds Corporation, Mississippi State University/USAID, and Rockefeller Foundation.

U.S. Agency for International Development, Bureau for Technical Assistance. 1973. *A Guide for Team Leaders in Technical Assistance Projects*. Washington, D.C.

APPENDIXES

Appendix A:
Guidelines for Research and
Basic Seed Activities

The guidelines that follow for the classification of cultivated plant populations were drafted by a committee of the Association of Official Seed Certifying Agencies. They have been slightly abridged.

After a thorough review of the constitution of present and anticipated products of plant breeding and of current classifications and descriptions, it is recommended that the various categories of cultivated varieties be described more precisely. The nature of variability of biological material makes development of "absolute" or "pure" definitions extremely difficult. Variability between certain of the categories is nearly continuous. Nevertheless, guidelines as developed here are considered essential for the orderly nomenclature and regulation of varieties.

Recognizing that scientific advances and discoveries have brought, and in the future may bring, about evolutionary changes in plant breeding methods and procedures, it is recommended that the following definitions be used as guidelines for classifying cultivated plant populations. Further advancements in the techniques of plant breeding may necessitate revisions of these guidelines in the future.

I. Variety (Cultivar)[1]

The term "variety" means a subdivision of a kind which is distinct, uniform, and stable: "distinct" in the sense that the variety can be differentiated by one or more identifiable morphological, physiological, or other characteristics from all other varieties of public knowledge; "uniform" in the sense that variations in essential and distinctive characteristics are describable; and "stable" in the sense that the variety will remain unchanged to a reasonable degree of reliability in its essential and distinctive characteristics and its uniformity when reproduced or reconstituted as required by the different categories of varieties.

A. Clonal varieties

Consist of one clone or several closely similar clones which are propagated by asexual means. Clonal varieties are propagated by such means as cuttings, tubers, corms, bulbs, rhizomes, divisions, grafts, or seed produced by obligate apomixis.

B. Line varieties

Consist of one or more lines of self- or cross-fertilizing plants and single line facultative apomicts having largely the same genetic background (a theoretical coefficient of parentage[2] of 0.87 or higher and 95 percent apomixis for the single line facultative apomicts; except in cases where it is not possible to achieve 95 percent apomixis, single line facultative apomicts with a level of apomixis as low as 80 percent may be classed as line varieties even though the variant plants present may vary in morphological characteristics) which are similar in essential and distinctive characteristics, and are maintained or reproduced by controlled self- or sib-fertilization or line crossing of the plants (for self- or cross-fertilizing plants) and by close generation control (for single line facultative apomicts).

C. Multiline varieties[3]

Consist of two or more near isogenic lines of normally self-fertilizing plants which are similar in most characteristics but differ in a limited number of describable physiological, morphological, or other essential or distinctive characteristics. A multiline is derived by growing the component lines separately and compositing the lines to constitute the breeder class of seed.

D. Open-pollinated varieties of cross-fertilizing crops

Consist of normally cross-fertilizing plants selected to a standard which may show variation but have one or more characteristics by which a variety can be differentiated from other varieties.

E. Synthetic varieties

1. First generation synthetic varieties (Syn-1)

Consist of first generation progenies derived by intercrossing a specific set of clones or seed propagated lines. These may include varieties of normally cross-fertilizing crops or of self-fertilizing crops into which mechanisms have been introduced to maximize cross-fertilization such as male sterility or self-incompatibility.

These varieties usually contain mixtures of seed resulting from cross-, self-, and sib-fertilization. The variety consists of only the first generation progenies after intercrossing and cannot be reproduced from seed of the first generation.

2. Advanced generation synthetic varieties (beyond Syn-1)

Consist of advanced generations derived from an initial intercrossing of a specific set of clones or seed propagated lines. Usually stable for only a limited number of generations.

F. Hybrid varieties (F_1)

Consist of first generation (F_1) progenies from a cross, produced through controlling the pollination, between (1) two inbred lines; (2) two single crosses; (3) a single cross and an inbred line; (4) an inbred line or a single cross and an open-pollinated or a synthetic variety; or (5) two selected clones, seed lines, varieties or species.

A line cross between two closely related inbreds (theoretical coefficient of parentage at least 0.87) is considered equivalent to a line (inbred) variety. The hybrid variety cannot be reproduced from seed of the hybrid generation.

G. F_2 Varieties

Consist of the next generation seed derived from the hybrid (F_1) generation. The variety cannot be perpetuated by growing additional generations.

II. Composite-Cross Populations

Consist of a population generation by hybridizing more than two varieties and/or lines of normally self-fertilizing plants and propagating successive generations of the segregating population in bulk in specific environments so that natural selection is the principal force acting to produce genetic change. Artificial selection also may be imposed on the population. The resulting population is expected to have a continuously changing genetic makeup. Breeder seed is not maintained as originally released.

III. Mixture[4]

Consists of seed of more than one kind or variety, each present in excess of 5 percent of the whole.

Notes

1. In these guidelines, the terms variety and cultivar are considered exact equivalents in accordance with the *International Code of Nomenclature of Cultivated Plants*.

2. Kempthorne, O. *An Introduction to Genetic Statistics*. Chapter 5. John Wiley and Sons, Inc., New York, 1957.

3. The committee voted four in favor and two opposed to classifying multilines as varieties. All other classifications were unanimous.

4. Blend is considered a synonym of mixture.

JOB SPECIFICATION FOR A SEED MAINTENANCE
AND MULTIPLICATION TECHNOLOGIST

Applicant should have adequate training and/or experience in agricultural botany, plant breeding, genetics, seed technology, and agronomy. Ability to describe and identify plant varieties is required with a knowledge of the morphological and physiological characteristics used for this work.

The main duties are to establish and supervise a maintenance and multiplication system for plant varieties. For each variety the technologist will follow these steps:

1. Establish head rows, plant rows, or other suitable plots from material supplied by the plant breeder using care to avoid mixture during planting.

2. Observe the characteristics of the growing plants, eliminate those not typical of the variety, and keep records on the results.

3. Supervise and participate in harvesting, threshing, drying, cleaning, packaging, and labeling of the separate bulk lots, and take steps to avoid admixtures.

4. Establish plots from the separate bulk lots from step 3.

5. Observe the characteristics of the growing plants and eliminate plants and/or all lots that are not typical of the variety.

6. Supervise harvesting, threshing, cleaning, packaging, and labeling of the bulked seed lots.

7. Arrange and supervise the multiplication of the bulked seed from step 6, and rogue as necessary.

8. Maintain details of all field observations and other records including the history, location, weight, and seed test results of each seed lot.

9. Ensure proper storage of each seed lot.

10. Maintain records on each seed lot including purity, moisture and germination percentages, and a complete inventory of all lots.

In addition, the technologist will maintain close liaison with the plant breeder responsible for the variety, assist in preparing descriptions of new varieties, and assist in training seed certification personnel and other colleagues in variety identification.

ALLOCATING BASIC SEED

A policy for the allocation of Basic Seed is important when (1) seed of a new variety is made available for the first time, (2) seed of established varieties is scarce, and (3) attempts are being made to encourage a group of seed growers and seed enterprises to participate in a seed certification program.

A committee can be formed to develop policies and monitor the application of the allocation system. The committee could have representation from the research station, the Basic Seed enterprise, the seed certifying authority, the farm advisory staff, and seed producing and marketing groups.

The following principles are suggested as a basis for developing an allocation policy:

1. Allocate seed to areas that trials have shown to be most suitable for the variety, if seed can be successfully produced there. (Cereal crops, for example, usually are grown for seed in the same areas where grain production takes place.) If the area where the variety is best adapted is not good for seed production, the seed should be allocated to growers in the most suitable seed-producing locations, and arrangements should be made for marketing the seed in the areas where the variety is adapted. (For example, many varieties of grasses and legumes produce seed better in drier areas but require a moist climate to give maximum forage yields.)

2. Within a chosen area, allocate seed to individuals and seed enterprises that have had experience in seed growing or to individuals and groups who are interested in seed production and marketing and are prepared to pay careful attention to detail, accept technical advice, and follow established standards. Participants in a program should have respect in their communities, be honest, and be dedicated to improving crop production.

3. Seed growers and seed enterprises accepting seed will make the entire production available for seed purposes. Individuals and organizations who accept seed shall be responsible for the processing, storage, and marketing of the production.

4. Choose seed growers and seed enterprises that are in locations that will minimize transport costs for seed technologists, who must visit the locations several times a season, and for moving seed from the farm to a processing facility.

5. Establish a minimum quantity for allocation. The minimum planting area should be large enough to keep lots separated on the farm at all times. Very small patches on a farm are easily damaged or destroyed by accident, isolation and inspection are difficult, and the harvested pro-

duce can be easily mixed accidentally.

6. Establish a maximum quantity for allocation. The area planted on any one holding should be within the capacity of the labor and equipment available. However, the size of an allocation should also be made after consideration of past seed production history.

7. Subject to the above considerations, allocate Basic Seed widely to minimize the risk of total failure, which might occur if all plantings are at a single location.

8. A list of persons *not* eligible should be developed based upon unfavorable past experience or perhaps upon official position held in the seed program or government.

A good system for allocating Basic Seed encourages seed production and marketing and stimulates interest in the program. Lack of a system or a system based upon favoritism undermines the program and discourages interest by capable seed growers.

Appendix B:
Three Case Histories

"TONGIL" RICE VARIETY*

Breeding rice varieties in the Republic of Korea is the responsibility of the Office of Rural Development (ORD), Suweon. In the summer of 1969, one line in the yield trials at Suweon caught the interest of the ORD director-general. It was IR667-98, a blast-resistant semidwarf that responded well to nitrogen and had long, heavy panicles and rather bold kernels of acceptable quality. Twelve kilograms of IR667-98 seed was flown to IRRI and multiplied during the winter of 1969-70. In April 1970, 600 kilograms of harvested seed were sent back to Korea.

IR667-98 did well in intensive local adaptation and cultural practice trials at ORD experiment stations across Korea that summer. Most of the seed was sent to IRRI for further multiplication during the winter of 1970-71, while Korean scientists multiplied seed in greenhouses. More than 5 tons of seed were shipped back to Korea. Together with locally produced seed, this was planted on 550 collective farms (5 hectares each) across the country in 1971. Rural guidance officers who had been trained at ORD crop experiment stations carefully supervised the growing of the new rice. To their satisfaction, it yielded an average of more than 7 t/ha. That year, ORD named IR667-98 as the Korean variety Tongil.

During the spring of 1972, Tongil was planted on 187,000 hectares across the country. . . . By 1975, Tongil covered a third of the rice-land—450,000 hectares. For the fourth consecutive year, Tongil yielded substantially more than the conventional varieties—7.0 vs 4.9 t/ha. Income on farms where Tongil was grown averaged $260 higher than on farms where conventional varieties were grown. . . .

A new variety called Yushin (IR1317-392-1/Tongil) has better eating quality than Tongil so it brings a higher price. Although its grains are more translucent and contain 21 percent amylose (2 percent lower than Tongil grains), they still cook somewhat drier than most Koreans prefer.

*Adapted from "How Tongil Triggered a Korean Rice Revolution," *IRRI Reporter* (1976).

Korea multiplied 100 tons of Yushin seed on 15 hectares of farmers' fields near IRRI during winter 1974-75. Three more tons of another new variety, Milyang 22 (IR1317-316/IR24), were grown in IRRI fields. The seed of both new varieties was flown back to Korea in time to plant during summer 1975.

ORD shipped to the Philippines 160 kilograms of seed of two newer varieties, Milyang 21 and Milyang 23 (also IR1317-316/IR24), during winter 1975-76. Korean scientists accompanied the seed and rented farmers' fields for multiplication. They harvested 60 tons of seed from 10.5 hectares and shipped the lot back in time to plant 300,000 hectares of Yushin and 1,600 hectares of Milyang in 1976.

TARAI DEVELOPMENT CORPORATION

Around 1960, several factors contributed to the later formation of the Tarai Development Corporation in India as well as to the start of the "green revolution" in Indian agriculture: (1) the start of stronger research programs on major crops—the all-India coordinated schemes; (2) the establishment of agricultural universities, which integrated teaching, research, and extension; (3) a greater involvement of researchers in the spread of technical know-how to farmers through a national demonstration program; and (4) a need and suitable situations for the development of a seed industry.

One of the universities formed was the G. B. Pant University of Agriculture and Technology. It was established in the Tarai Region of Uttar Pradesh in 1960 on a forty-four-hundred hectare farm. The farm had been used for seed production, but the university expanded the production capability. As the demand for seed increased, the university started involving progressive farmers in the Tarai in seed production.

In 1968 the government of India and the World Bank showed interest in strengthening the seed activity. The university prepared a draft project, "Integrated Agricultural Development Project, Tarai, U.P.," with production of good quality seed as the main feature. After a series of discussions, the project was approved in 1969. The Tarai Seed Development Project was started by setting up the Tarai Development Corporation Limited (TDC) as an autonomous company to implement the project. The project demanded good planning, implementation, and coordination by various groups: the G. B. Pant University of Agriculture and Technology, the National Seeds Corporation, the farmers of the Tarai area, the government of Uttar Pradesh, the government of India, the World Bank, the Uttar Pradesh State Electricity Board, the Exploratory Tube Wells Organization, the Agricultural Refinance Corporation, the State Bank of India, and the United Commercial Bank.

The TDC is a joint-sector enterprise. A thousand farmers hold two-fifths of the shares, the university has another two-fifths of the shares, and the National Seeds Corporation—an undertaking of the government of India—has the rest. The TDC is under the management of a board of directors, which has a proportionate representation of the three groups. The vice-chancellor of the university is ex officio chairman of the board.

The TDC selects growers from the farmer-shareholders to produce specific seed; arranges for Foundation Seed (Basic Seed) supplies and supervises the seed production program; manages and operates seed processing plants and related seed storage facilities; markets the processed seed; purchases farm machinery and other input supplies for project participants as necessary; arranges for the financing of various project activities; cooperates with financing banks to ensure proper loan repayments; assists project participants in farm development activities such as land leveling, construction of wells, irrigation systems, and drainage; and assists in arranging for increased electric power distribution systems to the area.

The university supports the project by participating in the all-India coordinated crop improvement programs and developing varieties, by operating an effective extension activity, by providing specialized services to project participants, by maintaining Breeder Seed of the varieties developed by the research scientists at the university, and by multiplying Foundation Seed of the varieties for which the university is responsible.

The project area, the Tarai, has several features suitable for seed production. It is a fertile belt of land, lying between the foothills of the Himalayas and the plains of northern India. Its natural slope minimizes the amount of land leveling necessary for efficient irrigation and drainage. It has abundant underground water resources. And there were few competing industries, a group of enterprising farmers, and a university that had already started to establish a reputation for supplying good quality seed.

The basic strategy in the project has been to develop a compact area and provide integrated development. It was recognized that the production of good quality seed would be possible only if the necessary inputs were made available. Moreover, each farm was helped to produce superior crops through farm mechanization, land development, irrigation development, electrification, adequate availability of fertilizers, and credit facilities.

The TDC concentrates on high-yielding varieties. The university and, in the initial period, the National Seeds Corporation provided a continuing supply of good quality Foundation Seed. The quantities of Certified

S~ed needed are carefully planned and allocations made to the university farm and participating farmers. Contracts are signed with each producing unit. A strict quality control program is followed by the TDC staff with some assistance from the university and the TDC attempts to provide seed that exceeds the Indian minimum seed certification standards. An official seed certification agency inspects the production. The TDC has gradually increased its seed processing capacity and now operates five processing plants in the project area. When seed arrives from the seed growers, great care is taken to process it promptly to maintain quality. A consistent pricing policy has been followed that allows for covering all costs, plus a suitable service charge to the corporation.

A strong marketing program was recognized from the beginning as being essential to the success of the TDC. Emphasis in the TDC's marketing program has been placed on assessing the effective demand for seed on which seed production is based, establishing an effective channel for distribution of seed by appointing distributors and dealers, arranging for proper storage for seed at strategic locations in selected areas, initiating activities to stimulate the demand for high quality seed, and offering post-sales service.

Both private and cooperative dealers have been appointed. Of the two thousand dealers and fifty-two distributors, the majority handle at least one other input, such as fertilizer.

In the mid-1970s, the program annually produced over thirty thousand tons of seed of rice, maize, millet, soybean, and wheat.

Although the TDC is unique in many ways, it illustrates a type of development that is possible when public and private resources are joined to achieve specific objectives. For more details on the TDC, write Tarai Development Corporation Limited, Pantnagar Naintal District, Uttar Pradesh, India.

KENYA SEED COMPANY

In 1956 a group of farmers formed the Kenya Seed Company Limited to meet the demand for seed of selections of indigenous grasses and clovers released by the Grasslands Research Station (now the National Agricultural Research Station) at Kitale. The Kenya Farmers Association supported the venture by taking up a large block of shares and agreeing to act as sole selling agent.

Although the founders and staff had as much knowledge as anyone about tropical grasses, they had much to learn. They knew that if they were going to use the grasses and legumes they had to master seed production techniques. Their early leadership gave the organization a good start.

But financial difficulties occurred, a common problem of such enterprises in their early years. In an attempt to reduce overhead costs by increasing volume, the Kenya Seed Company in 1958 began producing good quality, graded sunflower seed for sale on the international bird-feed market. Several other crops and sunflower varieties were tested before the decision was made to embark on the program. This step involved new machinery and extra storage space, but it was to be of unsuspected value to the company and the local farmers in the years to come.

Until 1963, the company was a relatively small enterprise, sometimes buoyant, often struggling, that catered to a limited market. The release of hybrids by the national maize program gave the company scope for expansion. The company was in a good position to add the new hybrids to its program because of the existing arrangements with growers of grass seed and because the machinery installed for sunflower processing had been originally designed for cleaning and grading maize. The company was granted sole production rights for the first hybrid, and the initial production of approximately forty hectares was undertaken in 1963.

During the next seven years the production of hybrid maize increased steadily. Concurrently new buildings and facilities were added: increased storage was followed by a modern processing plant exclusively for the hybrid maize, new offices, a laboratory, and more storage. The visible expansion of facilities went hand in hand with the not-so-obvious expansion in production and marketing. The distribution network became one of the most comprehensive in Africa. Simultaneously, the recruitment and training of staff were intensified. Now a complete, mobile team is continually in the field advising producers and customers and maintaining the distribution network.

As the demand for hybrid maize seed rapidly increased, the government asked the company to take over production of single crosses and, later, inbreds. The company found it necessary to produce this seed under its own control by renting small fields from growers near Kitale. This system had its disadvantages, and it was decided that land would have to be acquired if the necessary quantities of Basic Seed were to be produced to a sufficiently high standard. As a result, in 1969 an eight-hundred-hectare farm was leased from the government's Agricultural Development Corporation, which was, by this time, a substantial shareholder in the company. The farm produces the bulk of the annual Basic Seed requirements, serves as a demonstration unit, and provides facilities for the research and plant breeding section.

In the early 1970s, Kenyan officials became concerned about the state of the cereal seed supply. Under the existing system new varieties were rapidly contaminated by admixtures, and much of the seed offered for sale was of low quality. Consequently, at the request of the Ministry of Agriculture and the Wheat Board, the Kenya Seed Company began the production, processing, and distribution of certified wheat seed in 1972. A small office was opened in Nakuru, in the center of the wheat growing area, and storage and processing facilities were installed.

In 1974, after months of investigation and discussion, a new company called Seed Driers Limited was formed. It is a joint venture between the Kenya Seed Company and the Agricultural Development Corporation. Seed Driers Limited provides facilities for the artificial drying of the seed maize crop. The service saves a large amount of seed that otherwise would be spoiled by insects, diseases, and rodents as it dries in the field for up to four months. Another advantage is that artificial drying permits processing and distribution to be spread over a longer period of time. In addition, since the crop no longer must remain in the field to dry, the grower can cultivate his land early.

The company's annual seed production is over ten thousand tons of maize, four thousand tons of sunflower, thirty-five hundred tons of wheat, fifteen hundred tons of barley, one hundred fifty tons of grass, and eighty tons of bean.

Underlying the progress of the company is a spirit and determination that permeates the entire staff. The success of the enterprise is in part the result of this motivation, which is more difficult to obtain than a few hundred hectares of land or a new seed processing plant.

Appendix C:
Contract Information
and Sample Forms

SEED GROWER CONTRACT

Seed enterprises, Basic Seed enterprises, marketing groups, and government agencies often contract with seed growers to multiply seed. A contract or an agreement is usually signed between the enterprise or agency and the seed grower, outlining the conditions to be met and the obligations of both parties. These contracts vary considerably depending upon the crop; the responsibility to be assumed by each party; the purpose for which the seed is intended; and local, traditional, and legal expectations. It is not possible to provide a standard contract; however, the following points are often included.

1. Names and addresses of the parties to the agreement
2. The crop, variety, and generation or class of seed to be multiplied
3. The source and amount of the seed stocks to be used, the obligation for the costs involved in producing the seed crop, and the disposition of unused stocks
4. An indication of whether the production is to be certified and, if so, who is responsible for applying to the seed certifying authority and who will pay the fees
5. A statement of the total area to be planted and any restrictions about previous crops grown in the same field
6. An outline of general or specific cultural practices to be followed, such as weed control measures, plus special isolation requirements, and a statement of who bears the cost of special treatments
7. Who is to be responsible for and bear the costs of roguing to remove weeds and plants not characteristic of the variety, or to cover the costs of detasseling

8. A statement on special requirements during harvest (such as cleaning and inspection of equipment) and the desired seed moisture level for harvesting
9. Information on who is to store and insure the harvested seed, the place of delivery of the seed, and who is responsible for transportation and the costs involved
10. Who is responsible for seed drying, processing, and storage of cleaned seed and the costs incurred in these operations
11. The basis for payment: what special quality levels are to be met, whether the cleaned or uncleaned weights are to be used, the maximum moisture percentage on which the payment is to be calculated, and who is to receive the material for discard after cleaning
12. When payment may or must be made and the manner for determining the price to be paid including premiums, bonuses, or discounts in relation to seed quality
13. A statement on what is to be done with the produce if the field or the harvested crop fails to meet the requirements
14. The limits of liability if the seed crop is lost for reasons beyond the control of either party
15. The kind of information or special services that are to be provided by the seed enterprise or agency
16. The period or season for which the contract applies
17. A place for signature by both parties

Contracts can be simple and do not need to cover all the points listed. They are often divided into three sections to include obligations, responsibilities, and agreements by (1) the enterprise or agency, (2) the seed grower, and (3) both parties.

SEED ENTERPRISE RECORD FORMS

Each seed enterprise must have records and forms that fit the system it uses. Records are needed on individual seed lots as a part of having a well-organized and well-managed enterprise. In some countries seed legislation requires that records be kept. Such records usually include details on each seed lot purchased and received: its origin and how it is processed, treated, bulked, blended, packaged, stored, tested, labeled, sold, and shipped. A seed sample from the lot is also often part of the record. The forms shown—the Grower Receipt Form, the Seed Processing Plant Record, and the Warehouse and Sales Record—are provided as guides for the kind of information needed. The design and details of a form should be carefully evaluated to ensure that it fits the circumstances in which it is to be used.

Grower receipt form

GROWER RECEIPT FORM

No. _____   Date: _____

Grower's Name: _____

Address: _____

Crop: _____   Variety: _____

Class of Seed _____

Year Grown: _____   Field No.: _____

Number of Bags/Boxes: _____   Weight (gross): _____

Percent Moisture: _____   General Condition: _____

Estimated Cleanout: _____

Suggested Cleaning Sequence: _____

_____   _____

Grower   Warehouse Manager

Seed processing plant record form 2, page 1

Seed Processing Plant Record

RECORD NUMBER _____   INITIAL LOT NUMBER _____   FINAL LOT NUMBER _____

RECEIVING RECORD

Crop _____	Date	Number Units Received	Total net wt.	Receiving Report Number	Received by
Variety _____ Class _____					
Year _____ Field _____					
Grown _____ No. _____					
Grower _____					
Address _____					
Delivered Via _____					

DRYING RECORD

Date	Initial Moisture %	Drying Hrs.	Drying Temp.	Final Moisture %

PROCESSING RECORD

Date	Cleaning Time (hrs.)	Screens Used	Equip. Used	Clean Seed net wt.	Rejects net wt.	Air Loss net wt.	Net % Cleanout

TREATING RECORD

Date	Treating Material	Treating Rate	Net Wt.	Chemical Used

Seed processing plant record form 2, page 2

BAGGING RECORD

Date	Bag net wt.	No. of Bags	Type of Bags

RE-PROCESSING OR RE-BAGGING RECORD

Reason for: _____

Date	Operation	Wt. of Input Seed	Wt. of Output Seed

DELIVERY TO WAREHOUSE

Date	No. of Bags	Net Wt. per Bag	Total Wt.	Received by

QUALITY CONTROL RECORD

Kind of Sample	Date Sent / Received	Purity % Pure	Other crop	Weed Seed	Inert Matter	Germination % Normal	Abnormal*	Hard Seed	Dead*	Vigor* Rating	Moisture %	Laboratory No.	Initials
Pre-Processing													
After Processing													
Re-Test													
Re-Test													

Other Test Results _____

Details on Other Species Present:
Weed Seed _____
Crop Seed _____

* Information may not routinely be supplied by the Laboratory

Warehouse and sales record form

WAREHOUSE AND SALES RECORD

Record No.: _____

Lot. No.: _____

Crop: _____ Variety: _____

Class of Seed: _____

Year Grown: _____ Field No: _____

	Total Net Weight				Revalidation Record		
	No. Bags	Kg/Bag	Total Kg.		Date	Test No.	Germination %

Storage Location _____

Date	Sales or Shipping Receipt No.	Sold or Transferred to:	No. of Bags Sold	Net Wt. Sold	No. of Bags Remaining	Net Wt. Remaining	Method of Shipping	Remarks

SEED QUALITY CONTROL FORMS

Well-designed and convenient forms facilitate seed quality-control work and communication with the public. The number of forms should be held to a minimum, but they should be sufficient to meet the needs of the program.

Seed certification authorities normally have two basic forms—the Application Form for Seed Certification and the Field Inspection Report (forms 1 and 2). Field inspection report forms can normally be used for several crops, but special forms may be needed for certain crops.

Seed testing laboratories use forms mainly to (1) record samples received, (2) record details of test results within the laboratory, and (3) report test results to those who sent the samples for test (see forms 3, 4, and 5 for examples).

In seed law enforcement activities the primary requirements for forms normally include: (1) a report on individual seed lots inspected, (2) a stop-sale order, and (3) a report on the seeds tested. Forms 6 and 7 can serve for the first two purposes. The Seed Testing Laboratory Report (form 5) can serve as a test report form, with perhaps some modification of the statements at the bottom of the form. Depending on the exact legal requirements, it may be desirable, as a comparison, for this form to include the information claimed on the seed seller's label.

All of the forms provided in this section only indicate the kind of information needed. The requirements for a particular form and its design should be carefully evaluated to ensure that it is appropriate for the circumstances in which it is to be used.

Quality control form 1: Application for seed certification, page 1

APPLICATION FORM FOR SEED CERTIFICATION

Name of Applicant _____

District or Region _____

Address _____
       Street or Box Number   Town/Village

Telephone Number _____

State or Province _____

Variety	Class of Seed for Certification (Basic, Certified 1, Certified 2 or Certified 3)	Source of Seed	Evidence to Verify Seed Source Included*	No. of Fields**	Total Area	Fees***

TOTAL CERTIFICATION FEES: _____

Location of Farm and Fields

If this field is not located on the applicant's farm, provide information on where it is located:

Name _____ District _____

Address _____

What is the best way to reach the farm on which this field is located:
Nearest town, village and direction from villages _____

* Include all certification tags from bags of seed used or an invoice that shows the seed source and quantity of seed purchased and/or in the case of Basic Seed, a letter from the breeder indicating source of seed planted can be used.

** A field shall consist of an area which is occupied by a single class of a variety or hybrid and undivided, or if divided is separated by an intervening distance of not more than 50 meters.

*** Make check or money order payable to: Seed Certification Authority

Quality control form 1 (continued): Application for seed certification, page 2

FARM MAP AND LOCATION

N
|
W ——+—— E
|
S

Instructions

1. Show direction to nearest village or set of buildings where the seed certification technologist may obtain information or leave an inspection report

2. Show drainage ways, irrigation channels, roads, fields, buildings, and land marks

3. Indicate whether roads are dirt, gravel, tar or concrete

4. Number each field as 1, 2, 3, etc., and complete the following:

Field No.	No. of Ha. or Area	Variety	Class of Seed	Preceding Crop	Seeding Date
1					
2					
3					
4					
5					
6					

Date ————————————————

Signature of Applicant

252

Quality control form 2: Field inspection report (For hybrid crops this form should be modified to meet the crop's specific requirements)

FIELD INSPECTION REPORT

Name of Grower	Address	District & State of Province
Field No.	Class Being Inspected	Area in the Field
Crop and Variety	Source of Seed	Total Area of this Crop Inspected

Farm Location _____

COUNTS	FIRST INSPECTION		SECOND INSPECTION*	
	Varietal Mixture	Seed-borne Disease	Varietal Mixture	Seed-borne Disease
1				
2,				
3,				
4,				
5,				
6,				
7,				
8,				
9,				
10				
Average				
Percentage				
Varieties Present				
Name of Disease				

PRECEDING CROP: _____ ISOLATION: _____ ESTIMATED SEED YIELD: _____

INCIDENCE OF: DISEASES NOT SEED—BORNE: _____

INSECT PESTS: _____

NUMBER OF OTHER CROP PLANTS PER HECTARE _____

OBJECTIONABLE SEED BEARING WEEDS PRESENT: _____

REMARKS: _____

DOES THIS FIELD HAVE PROPER VARIETY CHARACTERISTICS? _____

DOES THIS FIELD MEET FIELD STANDARDS FOR CERTIFICATION? _____

SIGNATURE OF SEED GROWER OR HIS REPRESENTATIVE: _____

INSPECTED BY: _____ DATE: _____

* To be used if a second inspection is required or necessary

Quality control form 3: Seed laboratory receiving record

SEED LABORATORY RECEIVING RECORD

Laboratory Test No.	Date Sample Received	Date Sample Despatched to Laboratory	Sender's Name and Address	Crop and Variety	Class of Seed	Sample Designation or Lot Number	Quantity of Seed in Lot	Kind of Tests Required				Date Report Sent
								Purity	Germination	Moisture	Other	

Quality control form 4: Laboratory test results form, pages 1 and 2

LABORATORY TEST RESULTS FORM

Crop: Variety:		Class:	Sample Designation or Lot No.	Lab. Test No.

Report to:

Copies to:

Germination %	Hard Seed %	Pure Seed %	Other Crop Seed %	Weed Seed %	Inert Matter %	Moisture %	Notification

	First Count	Second Count	Total Normal Count	Abnormal	Hard Seed	Dead	Vigor Rating	Remarks
A								
B								
C								
D								
	% in Days	% in Days	Germination %	Abnormal %	Hard Seed %	Dead %		

General Remarks: Final Date:

LABORATORY TEST RESULTS FORM (Cont.)

Item	Grams	%	Grams	%	Mean %	Weed Seed	Grams	%	No. per kg.
Pure Seed									
Other Crop Seed									
Weed Seed									
Inert Matter									
Total									

Other crop seed	Grams	No. per kg.
Total		

Kind of Inert Matter

Analyzed by: Checked by:

Quality control form 5: Seed testing laboratory report

SEED TESTING LABORATORY REPORT

Address of Laboratory

Phone Number

To: Date: _____

Sampled by: _____

Crop: Variety:		Class:		Lot No.		Lab. Test No.
Germination %	Hard Seed %	Pure Seed %	Other Crop Seed %	Weed Seed %	Inert Matter %	Moisture %

Specific Seed Found

Other Crop Seed	Number per kg.	Weed Seed	Number per kg.

Other Test Results or Details _____

_____

_____

Summary Suggestions

Our laboratory has tested your seed and found it to be:

—— Good seed
—— Satisfactory seed but germination is low
—— Salable but containing excessive weed seed and needs further grading
—— Salable but containing excessive inert matter and needs further grading
—— Mislabeled with respect to crop
—— Too small for accurate analysis
—— Seriously infected with disease
—— Seriously infested with insects
—— Unsatisfactory and should not be used for seed

Quality control form 6: Seed inspection report, page 1

SEED INSPECTION REPORT

Date:_____

Seed Seller _____

Crop:_____

Address:_____

Variety:_____

Class:_____

Lot Number:_____

Number of
Container _____ Size _____ Total Quantity _____ Kind of Container _____ Number Sampled _____

┌─ LABELING INFORMATION ON CONTAINERS ─────────────────┐

Germination _____ %

Certification Labels ____ NO ____ YES

Hard Seed _____ %

Other Labels ____ NO ____ YES

Pure Seed _____ %

Imprinted on
Container ____ NO ____ YES

Crop Seed _____ %

Label Valid ____ NO ____ YES

Weed Seed _____ %

Treated ____ NO ____ YES

Inert Matter _____ %

Labeled as Treated ____ NO ____ YES

Specific Seeds Listed

Date of Labeling _____

Other Crop Seed	No./kg

Country or Region of Origin _____

_____

Weed Seed	

_____

Quality control form 6 (continued): Seed inspection report, page 2

PRESENT CONDITION

Seed appears properly labeled and sample taken	_____ NO	_____ YES
Seed unlabeled and requires labeling	_____ NO	_____ YES
Seed labeled but no longer valid	_____ NO	_____ YES
Seed treated but not labeled as treated	_____ NO	_____ YES
Seed improperly represented	_____ NO	_____ YES

ACTION TAKEN

Seller advised to :

Stop sales and correct labeling	_____ NO	_____ YES
Stop sales and correct advertising	_____ NO	_____ YES
Improve record system	_____ NO	_____ YES
Submit his own samples for testing more frequently	_____ NO	_____ YES
Study educational material left with him	_____ NO	_____ YES

Witness (If required) _____
Name

Seed Law Enforcement Technologist _____
Name

Quality control form 7: Stop-sale form

STOP - SALE FORM

DATE _____

Seller: _____

Address: _____

The following lots of seed are found to be in violation of the Seed Act Number _____

Crop	Variety	Lot No.	No. of Containers	Quantity	Kind of Violation	Sampled	
						Yes	No

By this notice you are ordered to hold this/these lot/lots of seed intact at:

_____

until compliance with the law has been achieved and the seed has been released from this order.

As soon as steps have been taken to remove the violation, please advise:

_____

Seed Law Enforcement Technologist

_____

Address and Telephone Number

Other Instructions _____

_____

Signed: _____

Seed Law Enforcement Technologist

Appendix D:
Model Seed Legislation

Possible wording for a General Seed Act and a Seed Certification Act is given to assist administrators and others who are involved in developing or changing legislation. These proposed acts should not be used without first assessing national needs and existing legislation.

GENERAL SEED ACT

Purpose: To regulate commerce in seed by preventing misrepresentation and to require certain standards with respect to named seed in order to provide users with seed suitable for their purpose.

1. *Definitions.** (1) "Minister" means the Minister of Agriculture or whomever he designates to act in his stead, (2) "Person" shall mean an individual, partnership, corporation, society, association, seed enterprise, trustee, receiver, or any government agency selling seed.
2. *Seed Regulated.* The provision of this Act shall regulate commerce in all agricultural, silvicultural, and horticultural seed for planting purposes as designated by the Minister in regulations (or decrees) promulgated by him or his agent.
3. *Harmful Weeds.* The kinds of harmful weed seed, and whether they shall be prohibited or limited in number in seed offered for sale, shall be determined by the Minister or his agent and published in regulations (or decrees).
4. *Quality Standards and/or Grades for Selling Seed.* Standards and/or grades for selling seed (such as minimum levels of percentages of pure seed, varietal purity, germination, germination and hard seeds or pure-live seed and maximum levels of weed seed) shall be established by the Minister or his agent by regulation (or decree).
5. *Labeling.* All seed subject to this Act as determined by the Minister shall be labeled when sold or offered for sale to show the following information:
   a. Name of the kind and variety

b. Lot identification
c. Origin, if determined by the Minister to be important
d. Percentage of pure seed
e. Percentage or number per unit of all weed seed
f. Kinds of harmful weed seed and the rate of each present as determined by the Minister
g. Percentage or number per unit of other crop seed
h. Percentage of inert matter
i. Percentage of germination of the pure seed
j. Percentage of hard or fresh, ungerminated seed, if present
k. Month and year the seed was tested
l. If treated with a toxic substance, the name and rate of the treatment and any caution statement required by the Minister or his agent
m. The name and address of the labeler or seller

[*or alternatively:*
5. *Labeling.* All seeds covered by this Act shall be labeled to show the following information:
   a. Name of the kind and variety
   b. Lot identification
   c. Grade, or the words "Complies with Seed Act Standards"
   d. Month and year of the germination test
   e. If treated with a toxic substance, the name and rate of the treatment and any caution statement required by the Minister or his agent
   f. Name and address of the labeler or seller]

6. *Imported Seed.* Seed imported from other states shall meet the standards established by the Minister for domestic seed.
7. *Registration of Seed Sellers.*[†] Any person who sells, or offers for sale, seed subject to this Act shall be required to register with the Ministry of Agriculture.
8. *Registration of Varieties.*[†] Any variety of seed sold, or offered for sale, must have been registered with the Ministry of Agriculture in accordance with regulations established by the Minister. No variety shall be deleted from the registered list without a three-year notice to allow disposal of all seed stocks in possession of vendors.
9. *False Advertising.*[†] It shall be unlawful for any person to disseminate, or cause to be disseminated, any false or misleading advertisement concerning seeds in any manner or by any means, including radio and television broadcasts, except that persons in the business of advertising and acting as agents for the advertiser shall not be held

liable for any advertising approved by the owner of the seed being advertised.

10. *Record Keeping.* All persons selling seed shall keep for a period of three years a complete record of seed bought and sold, as defined by the Minister. The Minister or his representatives shall have the right to inspect such records in enforcement of this Act. A sample of each lot of seed bought and sold shall be kept for one year after the lot it represents has been disposed of.

11. *Exemptions.* The provisions of this Act shall not apply to any common carrier transporting seed in the ordinary course of his business, *provided* that such carrier is not engaged in the processing or merchandizing of seed subject to this Act.

    The provisions of this Act shall not apply to seed produced by any grower on his own premises and sold directly to another grower, *provided* that the seller has not advertised such seed for sale, has not delivered such seed off his own premises by common carrier or otherwise, and is not in the business of buying and selling seed.

    The provisions of this Act shall not be deemed violated if seed is falsely represented as to variety or as to treatment with a toxic substance because the seed or treatment could not be identified because of indistinguishability in appearance from seed of the variety or treatment intended to be in the container or on the seed, respectively, provided that proper and complete records kept as provided in the regulations of the Minister disclose that the person selling or offering such seed for sale has taken all reasonable precautions to ensure the varietal identity or treatment of the seed to be that stated.

12. *Duties and Authority.* The duty of enforcing this Act and carrying out its provisions is vested with the Minister who may authorize agents to:
    a. Sample, inspect, make analysis of, and test all seed subject to the Act
    b. Enter public or private premises during regular workings hours in order to have access to seed and the records connected therewith subject to the Act and rules and regulations thereunder, and any truck or conveyor by land, water, or air at any time when the conveyor is accessible for the same purpose
    c. Have unfit seed seized and disposed of by order of a court of competent jurisdiction and/or issue and enforce a written or printed "stop-sale" of any lot of incompletely or falsely labeled seed subject to this Act that is found to be in violation thereof. Violation of a stop-sale order shall be punishable in the same manner as a violation of this Act.

13. *Sampling.* Sampling of seed to obtain a representative sample of a lot shall be in accordance with the rules for sampling established by the Minister.

14. *Seed Testing.* There are hereby authorized to be established such seed centers to test and inspect seed, subject to appropriations being made available, as may be necessary for carrying out the purpose of this Act. Seed designated by the Minister or his agent to be offered for sale shall be (officially) tested and supported by a record of such test as required by the Minister. The tests shall be made in conformance with the rules for seed testing as established by the Minister or his agent.

15. *Tolerances.* Tolerances between two or more tests, or between an official test and a claim on a label or a standard, made on the same lot of seed shall be recognized by the Minister or his agents before judging seed to be falsely represented.

[*or alternatively:*
15. *Tolerances.* The tolerances allowed between tests or between an official test and a claim on a label or standard shall be those established by the International Seed Testing Association before seed is judged to be falsely represented]

16. *Service Testing.* The Minister is authorized to provide tests for purity, germination, health, and other factors for any person and may establish reasonable fees for such testing.

17. *Exported Seed.* Seed offered for export may be sampled and tested for a fee by or for the Ministry of Agriculture before exportation. The Minister may establish minimum standards of quality for seed to be exported.

18. *Seed Certification.* The Minister may establish a seed certification system, as he deems necessary, to promote the use of seed with reasonable standards of genetic purity and other quality characters.[†]

19. *Unlawful Acts.* It shall be unlawful for any person to sell or offer for sale seed not in compliance with this Act.

[*or alternatively:*
19. *Unlawful Acts.* It shall be unlawful for any person to
   a. Sell or offer for sale seed subject to this Act
      1. Unless the germination test, required to be shown on the label, to determine whether the seed meets a designated standard has been made within _____ months prior to being sold or offered for sale, or unless the Minister has specified a shorter or longer

period for certain species or certain conditions§

2. Unless the seed is (i) labeled in accordance with this Act or (ii) meets the standards prescribed in the rules and regulations
3. If the labeling is false or misleading in any particular
4. If the seed contains excessive harmful weed seed as designated by the Minister in the rules and regulations
5. If the seed contains excessive weed seed as designated by the Minister in the rules and regulations
6. If the seed is labeled as a class of Certified Seed but does not meet the requirements of that class as defined by the Minister in the rules and regulations
7. If the seed is of a variety that has not been registered with the Ministry of Agriculture

b. Alter labeling or substitute seed in a manner that may defeat the purpose of this Act
c. Hinder or obstruct, in any way, any authorized agent in the performance of his duties under this Act
d. Fail to comply with a stop-sale order as provided in this Act
e. Sell seed treated with a toxic substance for other than seeding purposes, or sell seed not labeled as having been treated in accordance with this Act when such seed is sold for seeding purposes
f. Sell or offer for sale seed for seeding purposes unless the person so engaged is a registered seed seller as required by this Act
g. Fail to pay fees for service testing as provided for under this Act]

20. *Penalties.* The penalty for violation of this statute shall not be less than _____ for each offense nor more than _____. Minor violations may be disposed of by the Minister by the issuance of letters of warning, private hearings, or other administrative procedures established by the Minister.
21. *Appeals.* Appeals from decisions of the Minister or his agent may be made to a court of competent jurisdiction.
22. *Rules and Regulations.* The Minister is authorized to make such rules and regulations as he may deem necessary for the enforcement of this Act so long as such rules and regulations are consistent with the provisions and the intent of the Act, provided public notice and public hearings are carried out *before* rules and regulations are promulgated.
23. *Expenditures.* The Minister is authorized to make such expenditures as may be necessary for the enforcement of this Act and as may be appropriated for this purpose.
24. *Cooperation.* The Minister is authorized to cooperate with any of-

ficial department or agency or with any producing, trading, consuming, or scientific organization, domestic or international, in furthering administration of this Act.

25. *Employment, Education, and Research.* The Minister shall be authorized to employ such qualified personnel as funds appropriated allow for enforcement of this Act and to educate the public and conduct research and investigations with reference to seed.

26. *Delegation of Duties.* The Minister is authorized to delegate the duties devolving upon the Minister by this Act to be executed by such officers, agents, or employees of the Ministry of Agriculture as he shall designate for the purpose.

27. *Liability of Public Employees.* Employees of the Ministry of Agriculture shall not be held liable for any actions taken under this Act when such actions are authorized by the Minister and are in compliance with the Act and the rules and regulations thereunder.

28. This Act shall take effect as to those kinds of seed indicated by the Minister on a date specified by the Minister but beginning not later than _____ months following enactment.

Notes

*Terms that may need clarification should be selected from other sections of the completed draft of the General Seed Act and then defined in this section. See the Glossary for other suggested definitions.

†Including this section would depend upon the extent of seed law enforcement desired, as discussed in Chapter 5.

‡See Seed Certification Act below.

§The number of months must be specified according to the conditions prevailing in the country and the kind of packaging used.

SEED CERTIFICATION ACT

Purpose: To provide for certification as to the genetic purity and seed quality of prescribed kinds and varieties of seed.

1. Definitions (see the Glossary for some possible definitions)
2. The Minister shall, by establishing rules and regulations hereunder,
   a. Prescribe the manner in which persons may take part in seed certification
   b. Prescribe the kinds of seed and the varieties of seed that may be certified

    c. Prescribe the standards and requirements under which different classes of seed intended for certification shall be produced, processed, packaged, sealed, and labeled as Certified Seed

    d. Prescribe the records to be kept and the information required to be furnished by the persons who participate in the certification of seed

    e. Prescribe the fees to be paid

    f. Prescribe procedures for denial of participation by persons who violate regulations or fail to pay fees established by the Minister

    g. Prescribe for any other matter necessary for the successful implementation of this Act

3. All Certified Seed shall be labeled when sold or offered for sale to show (1) the class of Certified Seed, (2) the kind and variety of seed, (3) the lot identification, (4) conformity to minimum quality standards or the actual test figures, (5) the date of the germination test and/or the season for which the label is valid, (6) the name and rate of the treatment with a toxic substance, if any, and any caution statement required by the Minister or his agent, (7) the name of the issuing authority and the name, address, or code number of the Certified Seed producer.

4. *Expenditures.* The Minister is authorized to make such expenditures as may be necessary for the implementation of this Act and as may be appropriated or collected for this purpose.

5. *Cooperation.* The Minister is authorized to cooperate with any producing, trading, consuming, or scientific organization, domestic or international, in furthering implementation of this Act.

6. *Employment and Education.* The Minister shall be authorized to employ such qualified personnel as funds appropriated or collected for implementation of this Act allow and to educate the public with reference to Certified Seed.

7. *Delegation of Duties.* The Minister is authorized to delegate the duties devolving upon the Minister by this Act to an agency designated for this purpose or to such officers, agents, or employees of the Ministry of Agriculture as he shall designate for the purpose.

8. *Liability of Public Employees.* Employees of the Ministry of Agriculture or of any designated agent shall not be held liable for any actions taken under this Act when such actions are authorized by the Minister and are in compliance with the Act and the rules and regulations thereunder.

9. This Act shall take effect as to those kinds of seed indicated by the Minister on a date specified by the Minister but beginning not later than _____ months following enactment.

Appendix E:
International Seed Organizations

FEDERATION INTERNATIONALE DU COMMERCE DES SEMENCES

In many countries the seed enterprises have organized themselves into national seed industry associations. These associations may join the Fédération Internationale du Commerce des Semences (FIS), or the International Seed Trade Federation. A number of countries have no national seed industry association because their seed activities are concentrated in one enterprise or because the private seed industry has not developed to the extent that a national association can play a useful role. Over fifty countries are represented in FIS as regular or corresponding members. Under certain circumstances private firms may join FIS as corresponding members.

The major activities of FIS are (1) to form a liaison between the international seed industry and the international, governmental, and nongovernmental organizations; (2) to bring seed industry members from all over the world together through meetings; (3) to help create conditions for the international trade in seed so transactions can be completed smoothly; and (4) to give guidance and advice to members.

FIS was founded in 1924, a time when the international seed trade was confronting the need to standardize terms and procedures to facilitate transactions. Shortly after its formation, FIS worked closely with the newly organized International Seed Testing Association (ISTA) to develop a uniform method of seed analysis and a document for reporting the quality of a seed lot. Close cooperation between FIS and ISTA continues and the ISTA certificate is widely used by the seed industry.

FIS also drafted rules for the international trade in seed. After negotiations between seed traders of twelve countries, the first edition of the *Rules and Usages* for the international trade was approved in 1929.

Liaison between FIS and other international organizations interested in seed dates from after World War II. The major organizations include the Food and Agriculture Organization (FAO); the Organization for Economic Cooperation and Development (OECD); the European and

Mediterranean Plant Protection Organization (EPPO); and the Association Internationale des Selectionneurs pour la Protection des Obtentions Végétales (ASSINSEL), a nongovernmental international organization.

Every two years FIS organizes a congress to discuss association business and items of interest to the international seed industry. These meetings also offer seed traders an opportunity to meet colleagues from all over the world and, often, to begin negotiations. Each congress also makes it possible for members to exchange experiences and to discuss trends in the seed industry.

The contracts of seed enterprises doing international business usually mention that FIS rules will apply to the transaction. The rules were drawn up after careful study by all member-associations of FIS. The rules contain provisions on how transactions are to be completed and on other details related to quantity, shipment, insurance, packing, documents, payment, complaints, quality, and analyses.

Although the FIS rules minimize disputes, they sometimes may happen. Disagreements are normally resolved by a FIS arbitration committee on the basis of the *FIS Arbitration Procedure Rules for the International Seed Trade*. FIS, at the request of interested parties, occasionally approaches national governments regarding restrictive or other measures hampering the international seed trade.

FIS publishes a *Bulletin* in English, French, and German at irregular intervals, and each year the organization sends many circulars containing useful information to the member-associations and corresponding members.

FIS is financed by its member-associations and corresponding members who pay annual membership dues. As a rule, it does not give service to nonmembers.

The countries that have member-associations are:

Argentina	Germany (Fed. Rep.)	Spain
Australia	India	Sweden
Austria	Ireland	Switzerland
Belgium	Italy	Tunisia
Brazil	Japan	United Kingdom
Canada	Mexico	United States
Denmark	Morocco	of America
Finland	Netherlands	Venezuela
France	New Zealand	Yugoslavia
	Poland	

FIS headquarters can provide the addresses of member-associations in any country as well as information on joining.

Fédération Internationale du Commerce des Semences
Rokin 50, 1012 KV, Amsterdam, Netherlands

INDUSTRY COUNCIL FOR DEVELOPMENT

The Industry Council for Development (ICD) is a nonprofit organization, headquartered in New York, with worldwide operations. The ICD is financed by commercial enterprises in developed and developing countries without differentiation as to economic system or form of ownership. The ICD does not promote the interests of its members. Rather, it supports economic and social development by providing a channel for discussion and cooperation among decision makers in government, industry, and the United Nations system. The ICD has the support of the UN secretary-general and operates in cooperation with United Nations Development Programme and various other UN organizations. The ICD acts upon government requests that may be direct or through the UN system.

The ICD has established the Commercial Seeds Industry Development Project (CSIDP) to help improve the effectiveness of seed enterprises in developing countries. CSIDP operates within the broad framework of the ICD. Project activities are practical and results oriented. They help governments and individual seed enterprises overcome constraints such as a lack of specialized technology and knowledge, trained manpower, physical facilities, or financial partners. The activities also promote better understanding and recognition of the role of commercial seed enterprises whether they be public, private, or joint enterprises. There are three primary CSIDP activities:

Inventory of needs and resources. Operating on a worldwide basis, the project identifies the needs of developing countries and their seed enterprises, on the one hand, and, on the other, the resources available for commercial seed industry development, particularly from seed industries in developed countries as well as various institutions. This inventory of needs and resources is kept current through CSIDP cooperation with governments, seed industry associations, UN and other international organizations, universities, bilateral agencies, etc.

Promoting and assisting industrial cooperation. Using the inventory, the needs of developing countries are matched with identified resources. Cooperative arrangements involving appropriate seed organizations are then established. The objective of this "brokerage" function is to accelerate the transfer and adaptation of technological know-how, management skills, seed materials, marketing expertise, and capital equity where appropriate.

In the longer term, the project also promotes partnership arrangements between seed enterprises in all parts of the world, particularly those that will help developing countries achieve their food production goals. In all cases, the cooperation arranged by CSIDP is in accordance with the laws and policies of the host country.

Training. CSIDP can structure and help finance training programs for key personnel in various aspects of commercial seed operations. Individuals selected for CSIDP fellowships are generally already employed or scheduled for employment in technical or management positions of seed enterprises. Training programs are essentially on-the-job and, to the extent possible, take place at the facilities of cooperating seed enterprises. The exact content and length of each program vary with a trainee's needs but usually are planned to allow a trainee the opportunity to gain experience in a variety of operations during a seasonal cycle.

Additional information about the CSIDP is available from

Industry Council for Development
821 UN Plaza
New York, N.Y. 10017, U.S.A.

INTERNATIONAL SEED TESTING ASSOCIATION

The International Seed Testing Association (ISTA) was organized in 1924 to promote accurate and uniform methods of testing and evaluating seed. Through its activities it facilitates the efficient production, processing, distribution, and utilization of seeds, not only within countries, but also for seed moving in international trade.

ISTA has a worldwide membership accredited by governments of sixty countries. Approximately one hundred thirty official seed testing stations in these countries are authorized to issue ISTA's analysis certificates. Individual membership is restricted to persons who are engaged in the science or practice of seed testing or in technical control of these activities and who, in addition, are nominated by their governments. Member countries support ISTA through an annual membership fee, which is based upon the number of accredited stations in a country.

ISTA promotes uniformity in seed testing procedures through the widely used *International Rules for Seed Testing.* These rules are approved and amended from time to time at meetings of ISTA on the advice of its technical committees. Copies of the rules are available from ISTA in English, French, and German. Translations into Chinese, Portuguese, Russian, and Spanish have also been prepared in other countries.

Official seed testing stations authorized to do so may issue test results on special international analysis certificates, provided the tests have been carried out in accordance with the rules. The certificates are widely used in international trade and facilitate the movement of seed from one country to another.

To encourage the exchange of knowledge between seed technologists, ISTA publishes *Seed Science and Technology*. The *ISTA News Bulletin* is published three or four times a year to keep members and others informed about ISTA's activities. This publication is free. A complete list of publications available can be obtained from the association.

ISTA arranges triennial conventions and ordinary meetings. Papers are presented at the conventions and discussions are held on technical and scientific questions related to seed testing and seed research. The technical and scientific work of ISTA is the responsibility of special committees. Their reports are discussed and acted upon at the ordinary meetings. The exchange of seed samples, referee testing, and special training programs sponsored by ISTA contribute to uniformity in seed testing.

Although ISTA is a worldwide organization, many developing countries have not yet joined. The benefits to be gained from membership come primarily from the exchange of techniques and experiences with colleagues in other countries through ISTA's activities and publications. For information about publications and membership in ISTA write

International Seed Testing Association
P.O. Box 412, CH-8046, Zurich, Switzerland

INTERNATIONAL UNION FOR THE PROTECTION OF NEW VARIETIES OF PLANTS

The International Union for the Protection of New Varieties of Plants (UPOV) is an intergovernmental organization set up in 1961 by the International Convention for the Protection of New Varieties of Plants. The aims of the UPOV are (1) to promote and enhance plant breeding, thereby improving and developing agriculture, horticulture, and forestry, and (2) to harmonize the legal provisions in the member countries concerning plant variety protection in accordance with uniform and clearly defined principles. Initial membership was confined to a few countries in Europe where private plant breeding and an interest in achieving greater uniformity in protecting *privately* bred varieties existed, but interest in UPOV is growing in other countries where private crop breeding is important or is being encouraged. Developing countries with only public crop research have not participated in the program.

Two of the UPOV's most important tasks are to harmonize the procedures used in plant variety protection among its member countries and to simplify the methods used to examine new varieties for distinctness, homogeneity, and stability. As a result, *Guidelines for the Conduct of Tests for Distinctness, Homogeneity and Stability,* commonly referred to as Test Guidelines, have been prepared for most of the main crop species for which plant variety protection may be granted by the member countries. The guidelines are available in English, French, and German. The core of the guidelines is tables of characteristics used in making the examinations. Although seed programs in developing nations may not be interested in the UPOV program currently, the characteristics used for identifying varieties can be of value.

Considerable detail is used in describing and examining varieties for which plant variety protection is sought. Countries that are not trying to provide plant variety protection do not require the same amount of detail to describe varieties. They should not attempt to describe varieties in more detail than they need to meet their current objectives.

The UPOV has developed a "Model Agreement for International Cooperation in the Testing of Varieties." The model agreement can serve as a basis for bilateral agreements between two national offices of plant variety protection activities according to which one will, on request, perform tests on distinctness, homogeneity, and stability of varieties for the other or submit examination reports that already have been prepared, or are about to be prepared, to the other. Thus, one country can use another country's tests for granting plant variety protection or for other purposes, thereby saving labor and expense. Cooperation of this kind also promotes uniformity in the description of new varieties when they are being protected or released by more than one country. Possible changes in the UPOV program may broaden the membership and make the program of even greater value to the international commercial seed industry. For more information about publications and the program of UPOV write

International Union for the Protection of New Varieties of Plants
32, chemin des Colombettes, CH-1211 Geneva 20, Switzerland

OECD SCHEMES FOR SEED CERTIFICATION

The Organization for Economic Cooperation and Development (OECD) has established certification schemes for seed moving in international trade. Cereal seed (including rice), maize seed, herbage and oil seed, beet seed, seed of subterranean clover and similar species, vegetable seed, and forestry reproductive material are included in the OECD schemes.

Thirty-four countries participate in this activity. Because of the different conditions in the member countries, the OECD schemes establish standards only for varietal purity (except for the beet seed scheme, which also includes other seed quality standards). The OECD rules are available from its headquarters. The rules have application to seed certified under the scheme and intended for international trade. They do not directly affect the internal seed certification system of member countries. However, countries often work to harmonize their procedures so the internal and external systems are similar.

The following is adapted from various OECD publications (see Bibliography).

Methods of Operation

a. The Government of each country participating in the OECD Cereal Seed Scheme will designate the Authorities for the purpose of implementing the Scheme in that country.

b. The names and addresses of the Designated Authorities and any changes in their designation will be circulated by the OECD to all countries participating in the Scheme.

c. The operation and progress of the Scheme shall be reviewed at an Annual Meeting of representatives of the Designated Authorities. This Annual Meeting shall report on its work and make such proposals as may be deemed necessary to the Committee for Agriculture of the OECD.

d. The Annual Meeting shall each year nominate from amongst its members an Advisory Group. Its task shall be to advise the OECD Secretariat, when requested, on the technical aspects of the Scheme, to deal with urgent problems which may arise out of the implementation of the Scheme and to assist in the preparation of the next Annual Meeting.

e. The necessary co-ordination of the operation of the Scheme at the international level shall be ensured by the OECD.

f. When seed is labelled and sealed under one of the categories defined in these Rules and Directions, it is understood that all tests and inspections have been made in strict accordance with the Rules and Directions.

g. Certification and the use of the labels and certificates prescribed in these Rules and Directions shall not involve the OECD in any liability for compensation.

Definitions Used for Seed Classes:

Pre-basic Seed: Seed of generations preceding Basic Seed is known as Pre-basic Seed and may be of any generation between the parental material and the Basic Seed.

Basic Seed (Bred cultivars): The seed has been produced under the responsibility of the breeder according to the generally accepted practices for the maintenance of the cultivar and is intended for the production of certified seed.

Certified Seed: The seed is of direct descent from either Basic Seed or Certified Seed of a cultivar and is intended for the production of either Certified Seed or for purposes other than seed production.

Organization for Economic Cooperation and Development
2, rue André Pascal, 75775 Paris Cedex 16, France

Appendix F:
The Break-Even Chart

A break-even chart portrays the relationship of volume of sales or production to income and costs. By bringing together diverse facts in one comprehensive picture it aids managerial control. In a break-even chart, the horizontal axis gives volume in physical units (if a single product is involved) or value of sales (if many products are involved). Costs and income are plotted along the vertical axis. The income line is drawn from the lower left-hand corner of the chart. The total cost line is made up of indirect or direct costs.

A break-even chart makes it possible to see the effect of price change upon the volume required to break even and upon profit or loss. It shows the relative importance of the principal items of cost and how they vary with volume.

For example, with reference to the hypothetical enterprise illustrated, when the total cost line *BE* is composed of the increment of indirect cost *a* and direct cost *b*, management sees an opportunity to reduce the indirect cost items. This new indirect cost line would be represented by the line *KJ*. Since the direct cost is unchanged, the new total cost line becomes *KL*, and the break-even point is lowered from *C* to *G*, thus increasing profits or making it possible to earn a profit on a smaller volume of business. If, on the other hand, the direct costs increased, the total cost line would become *BI*, and the break-even point would move from *C* to *H*. If the selling price remained unchanged, this would mean that more goods would now have to be sold in order to break even. These are illustrative of the many types of analyses that can be made with the break-even chart.

The break-even chart enables management to anticipate the effects of its policies and decisions on the profitable operation of the enterprise as well as the effects of outside influences over which it has no control.

Simplified break-even chart

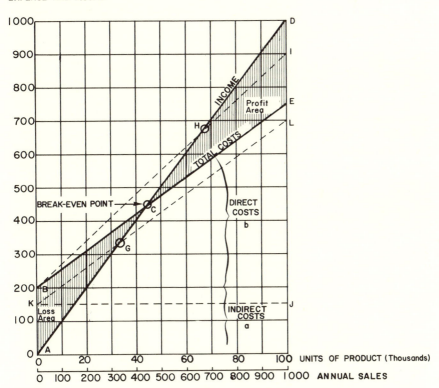

Source: U.S. Department of Agriculture. *Marketing: 1954 Yearbook of Agriculture.*
Washington, D.C.: Superintendent of Documents.

Appendix G:
Training in Seed Technology
in Developed Countries

Several universities and institutes in developed countries offer degree and nondegree programs with an emphasis on seed technology. The following is a list of some of the main locations and an indication of the kind of training at each place. For more information write the university or institute involved. Many other universities offer one or two special courses in seed technology or seed biology.

Denmark

Institute of Seed Pathology for Developing Countries, Ryvangs alle 78, DK-2900, Copenhagen, Hellerup, Denmark
> Offers training of six to twelve months in seed health testing. Training concentrates on practical, routine seed health testing of samples supplied by trainees. Lectures and special clinics on pathology and seed-borne diseases are included. The Danish International Development Agency grants scholarships to individuals from developing countries.

France

Centre de perfectionnement de l'institut national agronomique Paris-Grignon, 16, rue Claude Bernard, 75231 Paris Cedex 06, France
> Offers a course for persons having senior posts in seed firms or the seed industry.

Centre national d'études d'agronomie tropicale, 45, bis avenue de la Belle Gabrielle, 94130 Nogent sur Marne, France
> Offers a course to train people from developing countries in seed programs.

Groupe d'étude et de contrôle des variétés et des semences, INRA-GLSM, La Minière, 78000 Versailles, France

Arranges short or long training periods in various specialized laboratories for persons interested in seed testing at different levels (technicians, scientists, college graduates).

Netherlands

International Agricultural Centre, P.O. Box 88, Wageningen 6700 AB, Netherlands

Offers short (two to three months) postgraduate in-service training courses on vegetable seed technology, plant breeding, potato production, and other topics not directly related to seed. Participants (mainly from developing countries) must have at least a B.S. degree or the equivalent.

New Zealand

Seed Technology Centre, Massey University, Palmerston North, New Zealand

Assists developing countries, particularly in southeast Asia and the South Pacific, by offering short courses of four months duration; postgraduate diplomas in a course of approximately twelve months; M.S. and Ph.D. programs for a limited number of persons; and seminars and workshops in New Zealand and in the region.

United Kingdom

Edinburgh School of Agriculture, The King's Buildings, West Mains Road, Edinburgh EH9 3JG, United Kingdom

Offers postgraduate courses to prepare individuals for senior positions in organizations operating seed projects in developing countries: diploma in seed technology (nine months) and M.S. in seed technology (twelve months). It is possible for students seconded from industry to be accepted for one term and, thus, complete the program over a longer period of time.

United States of America

Seed Laboratory, Department of Botany and Plant Pathology, Iowa State University, Ames, Iowa 50010, U.S.A.

Offers courses in seed science and technology and provides practical experience through part-time employment.

Seed Technology Laboratory, Mississippi State University, P.O. Box 5267, Mississippi State, Mississippi 39762, U.S.A.

Offers programs with an emphasis in seed technology at the B.S., M.S., and Ph.D. levels. In addition, a two-month short course is conducted for people from developing countries and a one-week Short Course for Seedsmen is offered primarily for people in the United States. Consultants and assistance in seminars and workshops are also available.

Department of Farm Crops, Oregon State University, Corvallis, Oregon 97331, U.S.A.

Offers B.S., M.S., and Ph.D. degrees with a strong emphasis in seed technology.

Department of Agronomy and Soils, Washington State University, Pullman, Washington 99164, U.S.A.

Offers, through a joint arrangement with the University of Idaho, a wide range of courses on seed technology and production. Degree programs with a strong emphasis on seed technology are possible.

Department of Plant and Soil Sciences, University of Idaho, Moscow, Idaho 83843, U.S.A.

Offers courses in vegetable seed production, seed pathology, and seed physiology. A reciprocal agreement with Washington State University makes other seed technology courses available.

College of Agriculture, University of Arizona, Tucson, Arizona 85721, U.S.A.

Offers a B.S. degree in seed industry management.

Appendix H:
Useful Addresses

INTERNATIONAL AGRICULTURAL RESEARCH CENTERS

Following is a list of international research centers with crop interests that are associated with the Consultative Group on International Agricultural Research.

Asian Vegetable Research and Development Center (AVRDC)
P.O. Box 42, Shanhua, Tainan 741, Taiwan

Mung beans, soybeans, tomatoes, chinese cabbage, sweet potatoes

Centro Internacional de Agricultura Tropical (CIAT)
Apartado Aereo 6713, Cali, Colombia

Field beans, cassava, tropical forages, rice

Centro Internacional de Mejoramiento de Maíz y Trigo (CIMMYT)
Apartado Postal 6-641, Mexico 6, D.F., Mexico

Maize, wheat, barley

Centro Internacional de Papa (CIP)
Apartado 5969, Lima, Peru

Potatoes

International Board for Plant Genetic Resources, Crop Ecology and Genetic Resources Unit, FAO, Via delle Terme di Caracalla, 00100 Rome, Italy

International Centre for Agricultural Research in the Dry Areas (ICARDA)
P.O. Box 5466, Aleppo, Syrian Arab Republic

Barley, lentils, broad beans

International Crops Research Institute for the Semi-Arid Tropics (ICRISAT)
1-11-256 Begumpet, Hyderabad, 500016, A.P., India

Sorghum, millet, chick-peas, pigeon peas, groundnuts

International Institute of Tropical Agriculture (IITA)
P.M.B. 5320, Ibadan, Nigeria

Cowpeas, lima beans, sweet potatoes, yams, cassava

International Rice Research Institute (IRRI)
P.O. Box 933, Manila, Philippines

Rice

SEED PERIODICALS

A number of organizations and companies publish periodicals such as
newsletters, annuals, and magazines that are of interest to seed technologists and seed enterprises.

Abstract Services

Commonwealth Agricultural Bureaux, Farnham House, Farnham Royal,
Slough, SL2 3BN, United Kingdom (*Seed Abstracts*)

Associations

Australian Seed Producers Federation, 69 French St., Hamilton 3300,
Australia (*Australian Seed Producers Review*)

Department of Primary Industries, Meiers Road, Indooroopilly,
Queensland 4068, Australia (*Australian Seed Science Newsletter*)

Canadian Seed Growers' Association, Box 8455, Ottawa K1G 3T1 Ontario, Canada (*Seed Scoop*, issued quarterly)

Institute of Seed Pathology for Developing Countries, Ryvangs alle 78,
DK-2900, Copenhagen, Hellerup, Denmark (*Seed Pathology News*)

European and Mediterranean Plant Protection Organization, 1, rue le
Notre, 75016 Paris, France (Annual reports and specialized working-
group and panel reports; multilingual)

Organization for Economic Cooperation and Development, 2, rue André
Pascal, 75775 Paris Cedex 16, France (Publications on OECD certification schemes; multilingual)

Indian Society of Seed Technology, Division of Seed Technology, Indian

Agricultural Research Institute, New Delhi 110012, India (*Seed Research* and *Seed Tech News*)

Fédération Internationale du Commerce des Semences, Rokin 50, 1012 KV Amsterdam, Netherlands (*Newsletter*; multilingual)

International Seed Testing Association, P.O. Box 412, CH-8046, Zurich, Switzerland (*News Bulletin* and *Seed Science and Technology*; multilingual)

American Association of Seed Control Officials, Bureau of Plant Industry, Department of Agriculture, Harrisburg, Pennsylvania 17120, U.S.A. (Biennial proceedings)

American Seed Research Foundation, Suite 964, Executive Building, 1030 15th Street, N.W., Washington, D.C. 20005, U.S.A. (*Search*, published quarterly)

Association of Official Seed Analysts, c/o Charles Baskin, secretary, Mississippi State University, Box 5425, Mississippi State, Mississippi 39762, U.S.A. (*Newsletter*, published quarterly; *Proceedings*, published annually; *Journal of Seed Technology*; *Rules for Seed Testing*)

Association of Official Seed Certifying Agencies, Room C-227, P&AS Building, Clemson, South Carolina 29631, U.S.A. (*Certification Handbook* and *Annual Proceedings*)

Society of Commercial Seed Technologists, AMM Seed Testing, Box 1771, Fresno, California 93717, U.S.A. (*See Technologist News*, published quarterly; *Proceedings* of yearly conferences)

Commercial Magazines

Sedesem, 168, Bourse de Commerce, 75040 Paris Cedex 01, France (*Graines et Jardins*, published every two months)

Horfus Verlag GmbH, P.O. Box 550, 5300 Bonn-Bad Godesberg, Federal Republic of Germany (*Safa/Saatgutwirtschaft*, published monthly)

Zaadbelangen, Jan v, Nassaustraat 109, The Hague, Netherlands (*Zaadbelangen*, published monthly)

Seed World, 380 West Northwest Highway, Des Plaines, Illinois 60016, U.S.A. (*Seed World*, published monthly; *Seed Trade Buyers' Guide*, published annually)

Skarien and Associates, 1910 W. Olmos, San Antonio, Texas 78201, U.S.A. (*Seedsmen's Digest*, published monthly)

INFORMATION ON EQUIPMENT SOURCES

The following organizations and groups can supply current information about sources of drying, processing, testing, or small plot equipment in their country or, in some cases, internationally.

In addition, embassies can often provide information through their commercial attachés and other officials, and trade journals usually include many sources of supplies.

Empresa Brasileira de Pesquisa Agropecuaria (EMBRAPA), Ministerio de Agricultura, Caixa Postal 1316, 70.000 Brasília, DF, Brazil

Instituto Agronômico de Campinas, Caixa Postal 28, 13.100 Campinas, SP, Brazil

Engineering Research Service, Ottawa Research Station, Canada Department of Agriculture, Ottawa, Canada

Seed Section, Production and Marketing Branch, Canada Department of Agriculture, Ottawa, Canada

Seed Unit, Centro Internacional de Agricultura Tropical, Apartado Aereo 6713, Cali, Colombia

National Seeds Corporation, Ltd., Beej Bhavan, Pusa Complex, New Delhi 110012, India

Purchase and Procurement Branch, FAO, Via delle Terme di Caracalla, 00100 Rome, Italy

International Association on Mechanization of Field Experiments, Secretariate and Information Centre, LT1, 1432 Aas-NLH, Norway

Institute for Plant Breeding and Acclimatization in Radzikow, 05-870 Bfonie near Warsaw, Poland

Swedish Seed Testing and Certification Institute, S-171, 73, Solna, Sweden

University of Agriculture of Sweden, Department of Agricultural Technology, S-75007, Uppsala 7, Sweden

International Seed Testing Association, P.O. Box 412, CH-8046, Zurich, Switzerland

Official Seed Testing Station, National Institute of Agricultural Botany, Huntingdon Road, Cambridge CB3 OLE, United Kingdom

Edinburgh School of Agriculture, West Mains Road, Edinburgh, EH9 3JG, United Kingdom

National Institute of Agricultural Engineering, West Park, Silsoe, Bedfordshire, United Kingdom

Legume and Grass Seed Production Research Unit, Department of Agricultural Engineering, Oregon State University, Corvallis, Oregon 97331, U.S.A.

Seed Technology Laboratory, Mississippi State University, P.O. Box 5267, Mississippi State, Mississippi 39762, U.S.A.

Glossary

Definitions that have special relevance for use in seed laws are indicated by an asterisk (*).

AASCO. Association of American Seed Control Officials—an association of state and federal seed law enforcement officials in the United States.

Admixture. Something added to seed other than the kind and/or variety specified.

Advertisement. All representations other than those on the label, disseminated in any manner or by any means, related to seed within the scope of seed law.

Agronomic characters. Plant characteristics important in the adaptation of a plant to its commercial uses.

AOSA. Association of Official Seed Analysts—an association of government seed analysts of the United States and Canada.

AOSCA. Association of Official Seed Certifying Agencies—an association of seed certification agencies in the United States and Canada (known as the International Crop Improvement Association prior to 1968).

ASTA. American Seed Trade Association—an association of seed enterprises in the United States.

Basic Seed. A class of seed in a seed certification program that is the last step in the initial seed multiplications and is intended for the production of Certified Seed. (See Appendix E for the OECD definition.) Seed stock used for the same purpose as Basic Seed but not produced under a certification program is referred to as "the equivalent" of Basic Seed.

Breeder Seed. A class of seed in a seed certification program that is produced under the supervision of the plant breeder, originator, or owner of the variety; is controlled by that person or institution; and is the

285

source of the initial and recurring increases of Basic Seed.

Breeders' rights. The rights provided legally to a breeder, originator, or owner of a variety to control its production and marketing. The term is used synonymously with "plant variety protection."

Bulk. (1) Seed in nonpackaged form. (2) Seed of morphologically similar plants that has been mixed to form the base for multiplying a variety.

Cereal. Any grass species grown for its edible seed.

Certified Seed. (1) A class of seed that has been certified to conform to the standards for genetic purity established and enforced by a seed certifying authority; that is the direct descent of Breeder, Basic, or Certified Seed; and that is intended for the production of Certified Seed or for purposes other than seed production. (2) In a more general sense, certified seed is any class of seed—Breeder, Basic, or Certified—that has been handled so as to maintain satisfactory genetic purity and identity.

Commercial seed. Seed intended for crop production that has not been produced under a seed certification program.

Consumer. Any person who purchases or otherwise obtains seed for sowing but not for resale.

Cultural practices. All operations involved in the growing of crops, such as tillage, interrow cultivation, spraying, fertilization, and irrigation.

Dehumidifier. Equipment used to remove humidity from the air.

Detasseling. The removal by hand or machine of tassels of a female parent to prevent those plants from pollinating themselves during the production of hybrid maize seed.

Field inspection. An official inspection of a seed field. Normally associated with a seed certification program, but many "field inspections" are made outside of the seed certification activity.

FIS. Fédération Internationale du Commerce des Semences (see Appendix E).

Flat storage. Storage of seed in bags, in contrast to storage in bulk or nonpackaged form.

Genotype. The hereditary makeup of an individual plant that, with the environment, controls the plant's characteristics such as type of flower, shape of leaf, or color of seed.

Germplasm. The material basis of heredity. In a broad sense as used by plant breeders, it refers to seed and its heritable properties.

Grain. Seed consumed by humans or animals or processed for consumption.

* *Hermetically sealed container.* A metal container sealed by soldering, welding, or equivalent means to exclude all air movement; or, under U.S. law, a container that does not allow water vapor penetration through any wall, including the seals, greater than 0.05 grams of water per 24 hours per 100 square inches (645.2 square cm) of surface at 100°F (37.8°C) with a relative humidity of 90 percent on one side of the container wall and zero percent on the other.

Heterogeneous. Seed lots or plant populations that are not uniform, within established tolerances.

Homogeneous. Seed lots or plant populations that are uniform within established tolerances.

* *Hybrid.* The first-generation seed of a cross produced by controlling the pollination of and by combining (1) two or more inbred lines, (2) one inbred or a single cross with an open-pollinated variety, or (3) two varieties or species, except open-pollinated varieties of maize. The second generation or subsequent generations from such crosses shall not be regarded as hybrids. Hybrid designations shall be treated as variety names.

Inert matter. One of the components in a purity test conducted in a seed testing laboratory. It includes all nonseed material such as chaff, dirt, stones, and fungus bodies and seed material that is classified as inert under seed testing rules.

Interagency certification. The certification of seed through the cooperation of two seed certifying authorities or agencies. The process may involve many situations such as field inspection by one authority and final certification steps by another, the relabeling of Certified Seed after it has been rebagged, or issuing new labels on Certified Seed after an expiration date has passed if the seed has been found still to meet standards.

Isolation. A minimum distance needed by a seed multiplication field from other crops, varieties, or certain weeds to prevent contamination.

ISTA. International Seed Testing Association (see Appendix E).

Kind. A species, subspecies, or a group of species of crop plants known individually or collectively by one common name; for example, wheat, cotton, maize.

Label. Any tag, brand, mark, or other descriptive matter that is written, printed, stenciled, stamped, marked, embossed, or impressed on or attached to a seed container.

Labeling. Labeling includes all labels and other written, printed, or graphic representations, in any form whatsoever, accompanying or pertaining to any seed whether in bulk or in containers and includes representations on invoices.

Minister. The minister of agriculture or whomever he designates to act in his stead.

Mixture. Seed consisting of more than one kind of variety, each present in excess of 5 percent of the whole.

Nick. The proper synchronization of flowering between male and female parents in hybrid seed production.

OECD. Organization for Economic Cooperation and Development (see Appendix E).

Off-type. A plant or seed that deviates in one or more characteristics from that which has been described by the plant breeder or originator as being usual for the variety.

Parental material. A limited number of plants selected and used for the maintenance of a variety.

Percent germination. Percentage of seed that produces normal seedlings when tested in a laboratory according to established procedures.

Person. An individual, partnership, corporation, society, association, seed enterprise, trustee, receiver, or any government agency selling seed.

Plant variety protection. The protection provided legally to a breeder, originator, or owner of a variety to control its production and marketing. The term is used synonymously with "breeders' rights."

Population. As used in plant breeding, a population is a group of plants of the same species kept in a group to achieve certain breeding objectives.

Private hearing. A discussion of the facts between a seed law enforcement officer and a person charged with violating the seed law.

Processing. The cleaning, grading, scarifying, blending, or treating of seed. (Packaging and labeling of seed are also a part of the processing operation but may not be considered as "processing.")

Progeny. As used in plant breeding, the first generation of plants produced from a plant or plants.

Proprietary varieties. Varieties that are the sole property of an individual or an organization.

Pure live seed (PLS). The percentage of pure seed that germinates in a seed lot. Determined by multiplying the percentage of pure seed by the percentage of germination and dividing by 100.

Pure seed. The species stated by the sender or found to predominate in the test of a seed lot. (More precise definitions are included for each species in the seed testing rules of ISTA and many countries.)

Purity—analytical, physical, or mechanical. The percentage by weight of pure seed of a species as determined by a seed testing laboratory.

Purity—genetic or varietal. The purity with respect to variety as determined by a field inspection, by special laboratory tests when possible, or by inspection of plots planted for this purpose.

Range grasses. Grasses used for pasture or hay; they are often native to the area in which they are used.

Raw seed. Harvested seed that has not been cleaned and graded.

* *Record.* All information relating to a shipment or shipments of seed involved in transactions covered by a Seed Act and including a file sample of each lot of seed.

Relative humidity. Ratio of the quantity of water vapor actually in the air to the greatest amount possible at a given temperature.

Release. Making available to the public a variety for multiplication and use, or germplasm for use in breeding programs.

Rogue. (1) An off-type plant. (2) To remove an off-type plant.

Sample. Part of a seed lot presented for inspection or shown as evidence of the quality of the whole lot.

Sampling. The taking of seed from a seed lot for the purpose of official or unofficial quality assessment.

SCST. Society of Commercial Seed Technologists—an organization of seed analysts working in private and other nonofficial seed testing laboratories in the United States and Canada.

Sealed storage. Storage in a sealed container; usually refers to hermetic storage.

Seed. (1) A mature ovule consisting of an embryonic plant, a store of food, and a protective coat. (2) Parts of agricultural, silvicultural, and horticultural plants used for sowing or planting. Contrasted to grain, for example, which is used for consumption by humans and animals.

Seed enterprise. Any organization involved in seed growing, either directly or through contracts with others; drying; processing; storage; and marketing. It may or may not be involved in crop breeding research, and it may be a private or a government organization, or some combination.

Seed grower. An individual or institution that grows seed but does not process or market seed.

Seed increase. The multiplication of a quantity of seed by planting it to produce a larger quantity of seed.

Seed industry. The entire complex of organizations, institutions, and individuals associated with the seed program of a country. The *commercial* seed industry includes those individuals, seed enterprises, and marketing groups involved in producing and marketing seed for sale to consumers.

* Seed lot. A quantity of seed identified by a number or other mark, every portion or bag of which is uniform within recognized tolerances for the factors that are specified or appear on a label.

Seed marketing. The systematic determination of consumer needs, accumulation of seed and services to satisfy those needs, communication of information about the availability of seed and services, and distribution of seed to consumers.

Seed multiplier or producer. An individual who not only grows seed but also may process and sell seed, usually on a limited scale.

Seed program. The measures to be implemented and activities being carried out in a country to achieve the timely production and supply of seed of a prescribed quality in the quantities needed.

Seedstock. Seed that is used for multiplying additional cycles or generations of a variety. For example, it could be Basic Seed or seed that is equivalent to Basic Seed but is not a part of a seed certification system.

* *Seed testing terms.* The terms "pure seed," "germination," and other seed labeling and testing terms in common usage shall be defined as in *Seed Science and Technology* 4, no. 1, "International Rules for Seed Testing" of the International Seed Testing Association.

* *Seizure.* A legal process carried out by court order to dispose of a specific seed lot or lots alleged to be in violation of a law.

* *Stop-sale.* An administrative action based on authority provided in seed legislation to stop the sale of a seed lot that does not fulfill the legal requirements that are prescribed. If an infraction of the law is one that can be corrected, the stop-sale can be lifted after necessary corrective measures are taken and the seed lot may be sold.

* *Treated seed.* Seed that has received an application of a substance or has been subjected to a process for which a claim is made for seed treatment.

Trier. A device for sampling seeds in bags or in bulk.

UPOV. International Union for the Protection of New Varieties of Plants (see Appendix E).

Variant. Seed or plants that are distinct from a variety, occur in a predictable way, and were originally a part of the variety as released; thus they are not considered off-types.

Viability. Quality, such as the state of being alive; ability to live, grow, and develop, such as the viability of certain grains under dry conditions.

Vigor. Condition of good health and robustness; when seed is planted, vigor permits germination to proceed rapidly under a wide range of conditions.

* *Weed seed—harmful.* Seed of weeds that is found in crop seed, that is difficult to separate from crop seed during processing, and that is of weeds that are highly objectionable in fields because they are difficult to control or because of their undesirable effects on the crops produced.

Bibliography

This list contains publications grouped by language: English, French, Portuguese, and Spanish. Within each group, publishers or suppliers are listed alphabetically to facilitate ordering. Only publications currently available are included.

Publications in English

Academic Press, 111 Fifth Ave., New York, N.Y. 10003, U.S.A.
    Kozlowski, *Seed Biology*, vols. 1, 2, and 3 (1972)

Agricultural Development Council, 1290 Avenue of the Americas, New York, N.Y. 10019, U.S.A..
    Moseman, *Building Agricultural Research Systems in the Developing Nations* (1970)
    Mosher, *Getting Agriculture Moving* (1966)
    Mosher, *Serving Agriculture as an Administrator* (1975)

Agronomy Publications, P.O. Box 83, River Falls, Wisconsin, 54022, U.S.A.
    Delorit, *Illustrated Taxonomy Manual of Weed Seeds* (1970)

American Phytopathological Society, 3340 Pilot Knob Road, St. Paul, Minnesota 55121, U.S.A.
    Miller and Pollard, *Multilingual Compendium of Plant Diseases* (1976)

American Society of Agronomy, 677 South Segoe Rd., Madison, Wisconsin 53711, U.S.A.
    Hanson, *Alfalfa Science and Technology* (1972)
    Harlan, *Crops and Man* (1975)
    Sprague, *Corn and Corn Improvement*, 2d ed. (1977)

Association of American Seed Control Officials, Virginia Department of Agriculture, 1204 E. Main St., Richmond, Virginia 23219, U.S.A.
    The Seed Administrator's Handbook (1976)

Association of Official Seed Analysts, c/o Charles Baskin, secretary, Mississippi State University, Box 5425, Mississippi State, Mississippi 39762, U.S.A.
    Crosier, ed., *Rules for Testing Seeds* (1970)
    Grabe, ed., *Tetrazolium Testing Handbook for Agricultural Seeds* (1970)

Association of Official Seed Certifying Agencies, Room C-227, P&AS Building, Clemson, Scuth Carolina 29631, U.S.A.
  AOSCA Certification Handbook (1971)
  Hackleman, *History, 1919-1961 International Crop Improvement Association* (1961)

Avi Publishing Co., P.O. Box 831, Westport, Connecticut 06880, U.S.A.
  Brooker, Bakker-Arkema, and Hall, *Drying Cereal Grains* (1974)

Burgess Publishing Co., 7108 Ohms Lane, Minneapolis, Minnesota 55435, U.S.A.
  Copeland, *Principles of Seed Science and Technology* (1976)

Centre for Agricultural Publishing and Documentation, P.O. Box 4, Wageningen, Netherlands
  Bradnock, ed., *Advances in Research and Technology of Seeds*, Part I (1975)
  Thomson, ed., *Advances in Research and Technology of Seeds*, Part II (1976)
  _____, *Advances in Research and Technology of Seeds*, Part III (1978)
  _____, *Advances in Research and Technology of Seeds*, Part IV (1979)

Centro Internacional de Agricultura Tropical, Apartado Aereo 6713, Cali, Colombia
  Pinstrup-Andersen and Byrnes, eds., *Methods for Allocating Resources in Applied Agricultural Research in Latin America* CIAT/ADC Workshop, Cali, Colombia, November 26-29, 1974 (1975)

Centro Internacional de Mejoramiento de Maíz y Trigo, Apartado Postal 6-641, Mexico 6, D.F., Mexico
  Gerhart, *The Diffusion of Hybrid Maize in Western Kenya, Abridged by CIMMYT* (1975)

Chapman & Hall, Ltd., 11 New Fetter Lane, London, EC4 P 4EE, United Kingdom
  Roberts, ed., *Viability of Seeds* (1972)

Columbia University Press, 136 South Broadway, Irvington-on-Hudson, New York 10533, U.S.A.
  Barton, ed., *Bibliography of Seeds* (1967)

Commonwealth Agricultural Bureaux, Farnham House, Farnham Royal, Slough, SL2 3BN, United Kingdom
  Shaw and Bryan, eds., *Tropical Pasture Research—Principles and Methods* (1976)

Commonwealth Scientific and Industrial Research Organization, P.O. Box 89, East Melbourne, Victoria 3002, Australia
  Barnard, *Register of Australian Herbage Plant Cultivars* (1972)
  Ferns, Fitzsimmons, Martin, Simmonds, and Wrigley, *Australian Wheat Varieties: Identification According to Growth, Head, and Grain Characteristics* (1975)

Consultative Group on International Agricultural Research (Available from: United Nations Development Programme, One United Nations Plaza, New York, N.Y. 10017, U.S.A.)
International Research in Agriculture (1976)

Elsevier, P.O. Box 211, Amsterdam, Netherlands
Kahn, ed., *The Physiology and Biochemistry of Seed Dormancy and Germination* (1977)

FAO, Via delle Terme di Caracalla, 00100 Rome, Italy
FAO Seed Review 1974-75 (1977)
Farm Management Notes for Asia and the Far East 3, no. 2 (1967)
Feistritzer, ed., *Cereal Seed Technology* (1975)
Feistritzer and Kelly, eds., *Improved Seed Production* (1978)
Feistritzer and Redl, eds., *The Role of Seed Science and Technology in Agricultural Development* (1975)
Humphreys, *Tropical Pasture Seed Production* (1975)
Spitz, ed., *Case Studies in Seed Industry Development in Eight Selected Countries* (1975)

Fearon-Pitman Publishers, 6 Davis Dr., Belmont, California 94002, U.S.A.
Mager, *Preparing Instructional Objectives*, 2d ed. (1975)

Free Press (Available from: Macmillan Co., Riverside, New Jersey 08075, U.S.A.)
Rogers and Shoemaker, *Communication of Innovations*, 2d ed. (1971)

Illinois Crop Improvement Association, Inc., 508 S. Broadway St., Urbana, Illinois 61801, U.S.A.
Lang, *50 Years of Service: A History of Seed Certification in Illinois, 1922-1972* (1973)

Indian National Scientific Documentation Centre, CSIR Hillside Rd., New Delhi 110012, India (Available from: U.S. Department of Commerce, National Technical Information Service, Springfield, Virginia 22151, U.S.A.)
Saralidze and Khomeriki, *Machines for Extracting Tree Seeds* (1964)

Institut für Pflanzenbau und Saatgutforschung, Bundesallee 50, 3300 Braunschweig, Federal Republic of Germany
Seidewitz, *Thesaurus for the International Standardization of Genebank Documentation*, Part I—Cereals (1974); Part II—Forage Crops (1974); Part III—Root and Tuber Crops (1974); Part IV—Vegetables, Oil, and Fibre Crop Plants (1975); Part V—Selection of Common and Scientific Terms for Plant Pests and Diseases (1976)

Institute for Storage and Processing of Agricultural Produce, P.O. Box 18, Wageningen, Netherlands
Kreyger, *Drying and Storing Grains, Seeds, and Pulses in Temperate Climates*, Publikatie 205 (1972)

International Agricultural Development Service, 1133 Avenue of the Americas, New York, N.Y. 10036, U.S.A.
  Agricultural Assistance Sources, 2d ed. (1979)

International Association for Plant Taxonomy (Available from Bohn, Scheltema, and Holkema, P.O. Box 13079, 3507LB, Utrecht, Netherlands)
  Gilmour, *International Code of Nomenclature of Cultivated Plants* (1969)

International Association on Mechanization of Field Experiments, LT1, 1432 Aas-NLH, Norway
  The International Handbook on Mechanization of Field Experiments (1972)

International Development Research Centre, P.O. Box 8500, Ottawa K1G 3H9 Canada
  Nestel and MacIntyre, eds., *The International Exchange and Testing of Cassava Germ Plasm* (1975)

International Rice Research Institute, P.O. Box 933, Manila, Philippines
  Changes in Rice Farming in Selected Areas of Asia (1975)

International Seed Testing Association, P.O. Box 412, CH-8046, Zurich, Switzerland
  Proceedings of the International Seed Testing Association (Issues of special interest)
    "Number on Seed Legislation and Testing of Tropical and Subtropical Seeds" 36, no. 1 (1971)
    "OECD Standards, Schemes, and Guides Relating to Varietal Certification of Seed" 36, no. 3 (1971)
    "Seed Bibliography" 35, no. 4, and 36, no. 4 (1970 and 1971)
  Seed Science and Technology (Issues of special interest)
    "International Rules for Seed Testing" 4, no. 1 (1976)
    "Seed Cleaning and Processing" 5, no. 2 (1977)
  Survey of Equipment and Supplies (1973)
  Wellington, *Handbook for Seedling Evaluation* (1969)

Interstate Printers and Publishers, Inc., 19-27 N. Jackson St., Danville, Illinois 61832, U.S.A.
  Ware and McCollum, *Producing Vegetable Crops*, 2d ed. (1975)

Iowa Crop Improvement Association, 112 Agronomy Building, Ames, Iowa 50010, U.S.A.
  Robinson and Knott, *The Story of the Iowa Crop Improvement Association and Its Predecessors* (1963)

Iowa State University Press, Ames, Iowa 50010, U.S.A.
  Heath, Metcalfe, and Barnes, eds., *Forages: The Science of Grassland Agriculture*, 3d ed. (1973)

John Wiley & Sons, 605 Third Ave., New York, N.Y. 10016, U.S.A.
  Jugenheimer, *Corn: Improvement, Seed Production, and Uses* (1976)
  Knott, *Handbook for Vegetable Growers* (1957)

Mahlstede and Haber, *Plant Propagation* (1957)

Johns Hopkins University Press, Baltimore, Maryland 21218, U.S.A.
Wortman and Cummings, *To Feed This World: The Challenge and the Strategy* (1978)

Kansas Crop Improvement Association, Call Hall 205, Kansas State University, Manhattan, Kansas 66506, U.S.A.
Clapp, *The Kansas Seed Grower: A History of Seed Certification in Kansas, 1902-1970* (1970)

Lange & Springer, Heidelberg Platz 3, D-1000 Berlin 33, Federal Republic of Germany
Bewley and Black, eds., *Physiology and Biochemistry of Seeds*, Vol. 1 (1978)

Longman Group Ltd., Pinnacles, Harlow, Essex, CM20 2JE, United Kingdom
Bogdan, *Tropical Pasture and Fodder Plants (Grasses and Legumes)* (1977)
Humphreys, *Tropical Pastures and Fodder Crops* (1977)

McGraw-Hill Book Co., 1221 Avenue of the Americas, New York, N.Y. 10036, U.S.A.
Dahl and Hammond, *Market and Price Analysis: The Agricultural Industries* (1977)
Jenkins, *Modern Warehouse Management* (1968)

Macmillan Press Ltd., Little Essex Street, London WC2R 3LF, United Kingdom
Neergaard, *Seed Pathology*, Vols. 1 and 2 (1977)

Material Handling Institute, 1326 Freeport Rd., Pittsburgh, Pennsylvania 15238, U.S.A.
Apple, *Lesson Guide Outline on Material Handling Education—An Instructor's Guide* (1975)

Mississippi State University, Seed Technology Laboratory, P.O. Box 5267, Mississippi State, Mississippi 39762, U.S.A.
Delouche and Potts, *Seed Program Development* (1971)
Proceedings, 1976: Short Course for Seedsmen (1976)
(Also available are many articles and reprints on seed topics.)

National Seeds Corporation, Beej Bhavan, Pusa Complex, New Delhi 110012, India
A Handbook for Seed Inspectors (1972)

North Carolina State University, Office of International Programs, Raleigh, North Carolina 27607, U.S.A.
Bumgardner, Ellis, Lynton, Jung, and Rigney, *A Guide to Institution Building for Team Leaders of Technical Assistance Projects* (1971)

Organization for Economic Cooperation and Development, 2, rue André Pascal, 75775 Paris Cedex 16, France
Guide to Methods Used in Plot Tests and to Methods of Field Inspection of Cereal Seed Crops (1969)

Guide to the Methods Used in Plot Tests and to the Methods of Field Inspection of Herbage Seed Crops (1973)

OECD Scheme for the Varietal Certification of Cereal Seed Moving in International Trade (1977)

OECD Scheme for the Varietal Certification of Herbage and Oil Seed Moving in International Trade (1977)

OECD Scheme for the Varietal Certification of Maize Seed Moving in International Trade (1977)

OECD Scheme for the Varietal Certification of Subterranean Clover and Similar Species Moving in International Trade (1977)

OECD Scheme for the Varietal Certification of Sugar Beet and Fodder Beet Seed Moving in International Trade (1977)

OECD Scheme for the Varietal Certification of Vegetable Seed Moving in International Trade (1977)

Pennsylvania State University Press, 215 Wagner Building, University Park, Pennsylvania 16802, U.S.A.
Heydecker, ed., *Seed Ecology* (1973)

Pergamon Press, Elmsford, New York 10523, U.S.A.
Mayer and Poljakoff-Mayber, *The Germination of Seeds*, 2d ed. (1975)

Prentice Hall, Inc., Englewood Cliffs, New Jersey 07632, U.S.A.
Hartmann and Kester, *Plant Propagation: Principles and Practices*, 3d ed. (1975)
Still and Cundiff, *Essentials of Marketing*, 2d ed. (1976)
Tucker, *The Break-Even System; A Tool for Profit Planning* (1963)

RHM Arable Services, Throws, Stebbing, Dunmow, Essex, CM6 3AQ United Kingdom
Hervey-Murray, *A Preliminary Course in the Identification of Cereal Varieties* (1970)

Richard D. Irwin, Inc., 1818 Ridge Rd., Homewood, Illinois 60430, U.S.A.
McCarthy, *Basic Marketing: A Managerial Approach*, 6th ed. (1978)

Rockefeller Foundation, 1133 Avenue of the Americas, New York, N.Y. 10036, U.S.A.
Cummings, *Food Crops in the Low-Income Countries: The State of Present and Expected Agricultural Research and Technology* (1976)

Universiti Pertanian Malaysia, Serdang, Selangor, Malaysia
Chin, Enoch, and Harun, eds., *Seed Technology in the Tropics* (1977)

University of California Press, 2223 Fulton Street, Berkeley, California 94720, U.S.A.
Martin and Barkley, *Seed Identification Manual* (1973)

University of Minnesota Press, 2037 University Ave., S.E., Minneapolis, Minnesota 55455, U.S.A.

Christensen and Kaufmann, *Grain Storage—The Role of Fungi in Quality Loss* (1969)

University of the Philippines at Los Banos, Department of Development Communications, College, Laguna, Philippines
Jamias, ed., *Readings in Development Communication* (1975)

U.S. Agency for International Development, New Delhi (Order from: USAID, Washington, D.C. 20523, U.S.A.)
Gregg, Law, Virdi, and Balis, *Seed Processing* (1970)

U.S. Agency for International Development, Washington, D.C. 20523, U.S.A.
A Guide for Team Leaders in Technical Assistance Projects, PN AAB990 (1973)

U.S. Department of Agriculture (Available from: U.S. Government Printing Office, Washington, D.C. 20402, U.S.A.)
Justice and Bass, *Principles and Practices of Seed Storage,* Agricultural Handbook no. 506 (1978)
Reed, *Economically Important Foreign Weeds; Potential Problems in the United States,* Agricultural Handbook no. 498 (1977)
Schopmeyer, *Seeds of Woody Plants in the United States,* Agriculture Handbook no. 450 (1974)
Yearbook of Agriculture—Marketing (1954)
Yearbook of Agriculture—Seeds (1961)

Volunteers in Technical Assistance, 3706 Rhode Island Ave., Mt. Rainier, Maryland 20822, U.S.A.
Lindbland and Druben, *Small Farm Grain Storage* (1976)

W. H. Freeman & Co., 660 Market St., San Francisco, California 94104, U.S.A.
Heiser, *Seed to Civilization—The Story of Man's Food* (1973)
Janick, Schery, Woods, and Ruttan, *Plant Science—An Introduction to World Crops,* 2d ed. (1974)

Welsh Plant Breeding Station, Plas Gogerddan, Aberystwyth, Dyfed SY23 3EB, United Kingdom
Griffiths, Roberts, Lewvis, Stoddart, and Bean, *Principles of Herbage Seed Production* (1978)

West Publishing Co., P.O. Box 3526, St. Paul, Minnesota 55165, U.S.A.
Anderson, *Weed Science: Principles* (1977)

Publications in French

L'Association Canadienne des Producteurs de Semences, Case Postale 455, Ottawa, Ontario, Canada
Impuretés et éradication des plantes indésirables dans les céréales (n.d.)

Division de l'Information, Ministère de l'Agriculture du Canada, Ottawa, Ontario, Canada

Epuration des semences de céréales sur pied, Publication 1423 (1970)
Que savez-vous de la graine? Publication 1412 (1970)

FAO, Via delle Terme di Caracalla, 00100 Rome, Italie
Contrôle de la production et de la distribution des semences (1970)

Gauthier-Villars, 55, quai des Grands-Augustins, Paris (6è), France
Chaussat et Le Deunff, *La Germination des semences* (n.d.)

Institut National de la Recherche Agronomique, GEVES, La Minière, France
Liste alphabétique des principales espèces de plantes cultivées et de mauvaises herbes (n.d.)
Simon, *Les Maladies des céréales* (1971)

Association Internationale d'Essais de Semences, P.O. Box 412, CH-8046, Zurich, Suisse
Règles internationales pour les essais de semences; Règles et annexes 1976 (n.d.)

Publications in Portuguese

Associação Brasileira de Tecnologia de Sementes, a/c DNPV-DISEM, Edif. Palácio do Comércio, 70.000 Brasília, DF, Brasil
Anais do 1o. Encontro Nacional dos Técnicos em Análise de Sementes, 10 ENTAS (1971)

AGIPLAN, Edif. Venancio II, 70.000 Brasília, DF, Brasil
Associação de Analistas Oficials de Sementes, *Manual do teste de Tetrazólio* (n.d.)
Coletânea de artigos de Técnicos da AGIPLAN-M.A., *Tecnologia e produção de batatas-semente* (n.d.)
Delouche, *Pesquisa em sementes no Brasil* (1975)
Delouche & Potts, *Programa de sementes, planejamento e implantação* (1974)
Delouche, Still, Raspet, & Lienhard, *O Teste de tetrazólio para viabilidade da semente* (n.d.)
Fagundes & Gregg, *Manual de operações da Mesa de Gravidade* (1975)
Gregg, Vechi, Camargo, & Popinigis, *Roguing: sinônimo de pureza* (1974)
Gregg, Vechi, Lingerfelt, Camargo, & Popinigis, *Guia de inspeção de campos para produção de sementes* (1975)
Lingerfelt, *Padrões de campo para produção de sementes* (n.d.)
Musil, *Identificação de sementes de plantas cultivadas e silvestres* (n.d.)
Popinigis, *Fisiologia de sementes* (1974)
Popinigis & Rosal, *Colentânea de resumos de teses e dissertações sobre sementes* (n.d.)
Rocha, *Manual do teste de tetrazólio em sementes* (n.d.)
Vaughan, Gregg, & Delouche, *Beneficiamento e manuseio de sementes* (1976)
Welch, *Exercícios práticos sobre equipamentos de beneficiamento de sementes* (n.d.)

AGIPLAN (Abastecedor: Plano Nacional de Sementes, Ministério da Agricultura, Esplanada dos Ministérios, Bloco B, 70.000 Brasília, DF, Brasil)
Semente orgão técnico do PLANASEM—M.A. (n.d.)
Welch, *Beneficiamento de sementes no Brasil* (1973)

Departamento Nacional de Produção Vegetal, Divisão de Sementes e Mudas, Edifício Palácio do Comércio, 70.000 Brasília, DF, Brasil
Wetzel, *Lista bibliográfica de sementes* (n.d.)

Editora Agronômica "Ceres" Ltda., Caixa Postal 3917, São Paulo, Brasil
de Toledo & Filho, *Manual das sementes tecnologia da produção* (n.d.)

Editora Hucitec Ltda., Rua Beneficência Portuguesa, 44 1°s/105, 01033 São Paulo, SP, Brasil.
de Freitas, Filho, Aranha, & Bacchi, *Plantas invasoras de culturas no Estado de São Paulo*, Vol. 1 (1972); Vol. 2 (1975)

EMBRAPA, Ed. Super Center, Venâncio 2.000, Av. W/3 Sul Q. 700-BL. "B"-no 50, 70.000 Brasília, DF, Brasil
Popinigis, *Preservação da qualidade fisiológica da semente durante o armazenamento* (n.d.)

Ministerio da Agricultura, EPV-ETESEM (Abastecedor: Departamento Nacional de Produção Vegetal, Divisão de Sementes e Mudas, 70.000 Brasília, DF, Brasil)
Regras para análise de sementes (n.d.)

Publications in Spanish

Centro Regional de Ayuda Técnica-AID, México, D.F., México
Christensen & Kaufmann, *Contaminación por hongos en granos almacenados* (1976)
Departamento de Agricultura de los Estados Unidos, *"Semillas": Manual para el análisis de su calidad* (1965)

Compañía Editorial, Continental S.A., Calzada de Tlalpan Num. 4620, México 22, D.F., México
Departamento de Agricultura de los Estados Unidos, *Semillas: Anuario de agricultura* (1961)
Hartmann & Kester, *Propagación de plantas* (1971)
Ramirez Genel, *Almacenamiento y conservación de granos y semillas* (1966)

Ediciones Mundi-Prensa, Castelló 37, Madrid 1, España
Jean-Prost, *La Botánica y sus aplicaciones agrícolas* (1970)

Editorial Acribia, Apartado 466, Zaragoza, España
Berry et al., *Desecación y almacenamiento de granos* (1963)
Ede., *Producción de semillas pratenses* (1970)

FAO, Via delle Terme di Caracalla, 00100, Roma, Italia
Feistritzer, *Tecnología de la semilla de cereales* (1977)

Humphreys, *Producción de semillas pratenses tropicales* (1976)

Monro, *Manual de fumigación contra insectos*, 2d ed. (1970)

Instituto Colombiano Agropecuario, Tibaitatá, A.A. 151123, El Dorado, Bogotá, Colombia

Cárdenas, Reyes, & Doll, *Malezas tropicales*, Vol. 1 (1972)

Instituto Interamericano de Ciencias Agrícolas, San José, Costa Rica

León, *Fundamentos botánicos de los cultivos tropicales* (1968)

Productora Nacional de Semillas, México, D.F., México

Moreno Martinez, *Manual para el análisis de semillas* (1976)

Seed Technology Laboratory, Mississippi State University, P.O.. Box 5267, Mississippi State, Mississippi 39762, Estados Unidos de Norteamerica

Delouche, Still, Raspet & Lienhard, *Prueba de viabilidad de la semilla con Tetrazol* (1971)

Unión Tipográfica Editorial Hispano-Americana, Avenida de la Universidad 767, México 12, D.F., México

Miller, *Fisiología vegetal* (1967)